Zero Waste Lifestyle

Simple Steps to Minimize Waste and Maximize Environmental Impact

Felicia Bush

The presentation of the information is without contract or any type of guarantee assurance. The trademarks that are used are without any consent, and the publication of the trademark is without permission or backing by the trademark owner. All trademarks and brands within this book are for clarifying purposes only and are the owned by the owners themselves, not affiliated with this document.

Table of Contents

Chapter 1

Introduction to Zero Waste Living

Understanding the Zero Waste Philosophy

The Zero Waste Philosophy is more than just a trend; it's a transformative way of living that challenges the conventional norms of consumption and waste. At its core, this philosophy is about rethinking our relationship with the material world, recognizing the impact of our choices, and striving to minimize waste in every aspect of our lives. It is a call to action, urging individuals to become more conscious of their consumption patterns and to take responsibility for the waste they generate.

The journey towards a zero waste lifestyle begins with a fundamental shift in mindset. It requires an understanding that waste is not an inevitable byproduct of modern living, but rather a consequence of unsustainable practices. By embracing this philosophy, individuals commit to reducing their environmental footprint and contributing to a more sustainable future. This commitment is not about achieving perfection, but about making continuous improvements and striving for progress.

Central to the zero waste philosophy is the concept of the "5 R's": Refuse, Reduce, Reuse, Recycle, and Rot. These principles serve as a guide for making more sustainable choices and minimizing waste. Refusing unnecessary items, such as single-use plastics and disposable products, is the first step in reducing waste. By saying no to items that are not essential, individuals can prevent waste from being generated in the first place.

Reducing consumption is another key aspect of the zero waste philosophy. This involves being mindful of the products we purchase and the resources we consume. It means choosing quality over quantity, and prioritizing items that are durable, repairable, and made from sustainable materials. By reducing our consumption, we can decrease the demand for new products and the resources required to produce them.

Reusing items is an effective way to extend their lifespan and reduce waste. This can involve repurposing items for new uses, repairing broken items, or donating them to others who can use them. Reusing not only conserves resources but also encourages creativity and innovation in finding new ways to utilize existing items.

Recycling is an important component of the zero waste philosophy, but it should be seen as a last resort after refusing, reducing, and reusing. While recycling can help divert waste from landfills, it is not a perfect solution. Recycling processes require energy and resources, and not all materials can be recycled indefinitely. Therefore, it is crucial to prioritize the other R's before relying on recycling.

Rot, or composting, is the final principle of the zero waste philosophy. Composting organic waste, such as food scraps and yard waste, returns valuable nutrients to the soil and reduces the amount of waste sent to landfills. By composting, individuals can contribute to a healthier ecosystem and support sustainable agriculture.

The zero waste philosophy also emphasizes the importance of community and collaboration. Achieving a zero waste lifestyle is not an individual endeavor, but a collective effort that requires the support and participation of communities, businesses, and governments. By working together, we can create systems and infrastructures that facilitate waste reduction and promote sustainable practices.

Education and awareness are critical components of the zero waste philosophy. By educating ourselves and others about the impact of waste and the benefits of sustainable living, we can inspire change and encourage more people to adopt zero waste practices. This can involve sharing knowledge, resources, and experiences with others, as well as advocating for policies and initiatives that support waste reduction.

The zero waste philosophy also challenges us to reconsider our values and priorities. It encourages us to focus on experiences and relationships rather than material possessions, and to find fulfillment in living simply and sustainably. By shifting our focus away from consumerism, we can lead more meaningful and purposeful lives.

Adopting a zero waste lifestyle is not without its challenges. It requires effort, commitment, and a willingness to change long-standing habits. However, the rewards of living a zero waste lifestyle are significant. By reducing waste, we can conserve resources, protect the environment, and contribute to

a more sustainable future for ourselves and future
generations.

The Environmental Impact of Waste

Waste is an omnipresent issue that affects every
corner of our planet, from the depths of the oceans to
the peaks of the highest mountains. The
environmental impact of waste is profound and
multifaceted, influencing ecosystems, wildlife, and
human health in ways that are both visible and
insidious. Understanding these impacts is crucial for
anyone committed to reducing waste and living more
sustainably.

One of the most immediate and visible impacts of
waste is pollution. Landfills, often the final resting
place for much of our waste, are not just unsightly;
they are also significant sources of pollution. As waste
decomposes, it releases methane, a potent greenhouse
gas that contributes to climate change. Methane is
over 25 times more effective at trapping heat in the
atmosphere than carbon dioxide over a 100-year
period, making landfills a major contributor to global
warming. Additionally, leachate, a toxic liquid formed
when rainwater filters through waste, can seep into
the ground and contaminate soil and water sources,
posing a threat to both ecosystems and human health.

The oceans, too, bear the brunt of our wasteful habits.
Every year, millions of tons of plastic waste end up in
the oceans, where they wreak havoc on marine life.
Sea turtles mistake plastic bags for jellyfish, ingesting

them with fatal consequences. Birds and fish become entangled in discarded fishing nets and six-pack rings, leading to injury or death. Microplastics, tiny fragments of plastic that result from the breakdown of larger items, have infiltrated the marine food chain, affecting organisms from plankton to whales. These microplastics are not only ingested by marine animals but also by humans, as they make their way into the seafood we consume.

The impact of waste extends beyond pollution and into the realm of resource depletion. The production of goods that eventually become waste requires vast amounts of natural resources, including water, energy, and raw materials. For example, the production of a single plastic bottle requires significant amounts of petroleum, a non-renewable resource. When these items are discarded rather than recycled or reused, the resources used in their production are effectively lost. This unsustainable cycle of production and disposal places immense pressure on the planet's finite resources, contributing to deforestation, habitat destruction, and biodiversity loss.

E-waste, or electronic waste, is another growing concern. As technology advances at a rapid pace, electronic devices become obsolete quickly, leading to a surge in discarded gadgets. E-waste contains hazardous materials such as lead, mercury, and cadmium, which can leach into the environment if not disposed of properly. These toxic substances pose serious health risks to humans and wildlife, contaminating soil and water and entering the food chain. Moreover, e-waste often ends up in developing countries, where informal recycling practices expose workers to dangerous conditions and further exacerbate environmental degradation.

The environmental impact of waste is not limited to pollution and resource depletion; it also affects climate change. The production, transportation, and disposal of goods contribute to greenhouse gas emissions at every stage of their lifecycle. From the energy-intensive processes of manufacturing to the emissions from transporting goods across the globe, waste is intricately linked to the carbon footprint of human activity. By reducing waste, we can significantly decrease our carbon emissions and mitigate the effects of climate change.

Addressing the environmental impact of waste requires a multifaceted approach that involves individuals, communities, businesses, and governments. Individuals can make a difference by adopting sustainable practices such as reducing consumption, reusing items, and recycling responsibly. Communities can implement waste reduction programs and promote awareness about the importance of sustainable living. Businesses can innovate by designing products with sustainability in mind, using eco-friendly materials, and implementing circular economy principles. Governments can enact policies and regulations that incentivize waste reduction and support sustainable practices.

Education and awareness are key components in tackling the environmental impact of waste. By understanding the consequences of our wasteful habits, we can make informed choices that align with our values and contribute to a healthier planet. This involves not only educating ourselves but also sharing knowledge with others and advocating for systemic change.

Benefits of Adopting a Zero Waste Lifestyle

Embracing a zero waste lifestyle offers a multitude of benefits that extend beyond the individual to the community and the planet as a whole. This approach to living is not merely about reducing waste; it is a holistic transformation that can lead to a more fulfilling, sustainable, and conscious way of life. By adopting zero waste practices, individuals can experience personal growth, financial savings, and a deeper connection to the environment.

One of the most immediate benefits of a zero waste lifestyle is the potential for significant financial savings. By focusing on reducing consumption and reusing items, individuals can cut down on unnecessary purchases and avoid the costs associated with disposable products. For instance, investing in reusable items such as water bottles, coffee cups, and shopping bags can eliminate the need for single-use alternatives, leading to long-term savings. Additionally, by embracing practices like meal planning and bulk buying, individuals can reduce food waste and save money on groceries. These financial benefits are not only advantageous for individuals but also contribute to a more sustainable economy by reducing demand for disposable goods.

Beyond financial savings, a zero waste lifestyle encourages a shift towards minimalism and intentional living. By prioritizing quality over quantity, individuals can declutter their lives and focus on what truly matters. This shift can lead to a greater sense of fulfillment and contentment, as individuals learn to appreciate experiences and relationships rather than material possessions. The process of simplifying one's life can also reduce stress and increase mental clarity, as individuals are no longer burdened by the constant pursuit of consumer goods.

The environmental benefits of a zero waste lifestyle are profound and far-reaching. By reducing waste, individuals can decrease their carbon footprint and contribute to the fight against climate change. The production and disposal of goods are significant sources of greenhouse gas emissions, and by minimizing waste, individuals can help mitigate these impacts. Furthermore, reducing waste conserves natural resources, such as water, energy, and raw materials, which are often depleted in the production of disposable goods. By conserving these resources, individuals can help protect ecosystems and biodiversity, ensuring a healthier planet for future generations.

A zero waste lifestyle also fosters a deeper connection to the environment and a greater awareness of the impact of individual actions. By becoming more conscious of waste and consumption, individuals can develop a sense of stewardship and responsibility for the planet. This heightened awareness can lead to more sustainable choices in other areas of life, such as transportation, energy use, and food consumption. As individuals become more attuned to the natural world, they may also experience a greater sense of wonder and appreciation for the beauty and complexity of the environment.

Community engagement is another significant benefit of adopting a zero waste lifestyle. By participating in local initiatives, such as community gardens, composting programs, and waste reduction workshops, individuals can connect with like-minded people and build a sense of community. These connections can lead to the sharing of resources, knowledge, and support, creating a network of individuals committed to sustainable living. Additionally, by advocating for zero waste practices and policies, individuals can contribute to positive change at the community level, influencing businesses and governments to adopt more sustainable practices.

The health benefits of a zero waste lifestyle should not be overlooked. By reducing exposure to harmful chemicals found in many disposable products, such as plastics and synthetic materials, individuals can improve their overall health and well-being. For example, by choosing natural and organic personal care products, individuals can avoid the potential health risks associated with synthetic chemicals and additives. Additionally, by reducing food waste and focusing on fresh, whole foods, individuals can improve their diet and nutrition, leading to better health outcomes.

Adopting a zero waste lifestyle can also inspire creativity and innovation. By finding new ways to repurpose and reuse items, individuals can tap into their creative potential and develop new skills. This creativity can extend to other areas of life, such as cooking, gardening, and crafting, leading to a more enriched and fulfilling life. The process of finding innovative solutions to reduce waste can also foster a sense of empowerment and self-sufficiency, as individuals learn to rely on their resourcefulness rather than consumer goods.

Common Misconceptions About Zero Waste

The concept of zero waste often conjures images of a life stripped of convenience, filled with endless sacrifices and unattainable ideals. However, these perceptions are largely based on misconceptions that can deter individuals from embracing this transformative lifestyle. By addressing these misunderstandings, we can demystify zero waste and reveal it as an accessible and rewarding path towards sustainability.

One prevalent misconception is that zero waste living requires perfection. Many people believe that to be truly zero waste, one must produce no waste at all, which can seem daunting and unrealistic. In reality, zero waste is not about achieving absolute zero but about making a conscious effort to reduce waste as much as possible. It's a journey of continuous improvement, where small, incremental changes can lead to significant impacts over time. By focusing on progress rather than perfection, individuals can alleviate the pressure and embrace the zero waste lifestyle with a more positive and realistic mindset.

Another common myth is that zero waste living is expensive. The perception that sustainable products and practices come with a hefty price tag can be a significant barrier for many. However, zero waste living often leads to financial savings in the long run. By reducing consumption, reusing items, and prioritizing quality over quantity, individuals can cut down on unnecessary expenses. For example, investing in durable, reusable items such as stainless steel water bottles or cloth shopping bags can eliminate the need for disposable alternatives, saving money over time. Additionally, practices like meal planning and bulk buying can reduce food waste and lower grocery bills. While some initial investments may be required, the overall financial benefits of zero waste living can be substantial.

A third misconception is that zero waste living is time-consuming and inconvenient. The idea of making everything from scratch, carrying reusable containers everywhere, and meticulously sorting waste can seem overwhelming. However, zero waste living is about finding a balance that works for each individual. It doesn't require a complete overhaul of one's lifestyle but rather the integration of sustainable practices into daily routines. Simple changes, such as carrying a reusable water bottle or using cloth napkins, can become second nature with time. Moreover, many zero waste practices, like meal prepping or buying in bulk, can actually save time and simplify life in the long run.

Some people believe that zero waste living is only for environmentalists or those with a particular lifestyle. This misconception can create a sense of exclusivity, deterring individuals who don't identify with these labels. In truth, zero waste is for everyone, regardless of background, lifestyle, or beliefs. It's about making conscious choices that align with one's values and circumstances. Whether it's a busy professional, a student, or a family, anyone can adopt zero waste practices that fit their unique situation. The key is to start small and build on successes, creating a personalized approach to sustainability.

Another myth is that zero waste living is ineffective or insignificant in the grand scheme of things. Some individuals may feel that their efforts are just a drop in the ocean, unable to make a meaningful difference. However, the collective impact of individual actions can be powerful. By reducing waste, individuals contribute to a larger movement that drives demand for sustainable products, influences policy changes, and inspires others to take action. Every small step towards zero waste is a step towards a more sustainable future, and the ripple effect of these actions can lead to significant change.

There is also a misconception that zero waste living is rigid and restrictive, leaving no room for flexibility or enjoyment. This perception can make the lifestyle seem unappealing and unattainable. In reality, zero waste living is about creativity and adaptability. It's about finding joy in simplicity, discovering new ways to repurpose items, and exploring sustainable alternatives. The journey towards zero waste can be an opportunity for personal growth, creativity, and connection with others who share similar values. By approaching zero waste with an open mind and a sense of curiosity, individuals can find fulfillment and satisfaction in their sustainable choices.

Lastly, some people believe that zero waste living requires giving up all modern conveniences and returning to a primitive way of life. This misconception can create fear and resistance to change. However, zero waste is not about rejecting modernity but about reimagining it in a more sustainable way. It's about embracing innovation and technology that support sustainable practices, such as energy-efficient appliances, digital solutions to reduce paper waste, and eco-friendly materials. By integrating these advancements into daily life, individuals can enjoy the benefits of modern living while minimizing their environmental impact.

Setting Personal Goals for Waste Reduction

Embarking on a journey towards waste reduction begins with setting personal goals that are both realistic and inspiring. These goals serve as a roadmap, guiding individuals through the myriad of choices and changes that come with adopting a more sustainable lifestyle. By setting clear and achievable objectives, individuals can maintain motivation, track progress, and celebrate successes along the way.

The first step in setting personal goals for waste reduction is to assess your current habits and identify areas for improvement. This involves taking a close look at your daily routines and consumption patterns to understand where waste is being generated.

Consider keeping a waste diary for a week, noting down every item you discard and the circumstances surrounding it. This exercise can provide valuable insights into your waste habits and highlight specific areas where changes can be made.

Once you have a clear understanding of your current waste footprint, it's time to set specific, measurable, achievable, relevant, and time-bound (SMART) goals. These goals should be tailored to your unique lifestyle and circumstances, ensuring they are both challenging and attainable. For example, if you notice that a significant portion of your waste comes from single-use plastics, a SMART goal might be to reduce plastic waste by 50% within three months by switching to reusable alternatives.

Breaking down larger goals into smaller, manageable steps can make the process less overwhelming and more achievable. For instance, if your overarching goal is to reduce household waste by 30% in six months, you might start by focusing on one area at a time, such as the kitchen or bathroom. By tackling one aspect of waste reduction at a time, you can build momentum and confidence as you progress towards your larger goal.

Incorporating waste reduction goals into your daily routine is essential for long-term success. This might involve setting reminders to bring reusable bags when shopping, planning meals to minimize food waste, or dedicating time each week to composting. By integrating these practices into your everyday life, they become second nature, reducing the likelihood of reverting to old habits.

Accountability is a powerful tool in achieving personal goals. Sharing your objectives with friends, family, or a community of like-minded individuals can provide support, encouragement, and motivation. Consider joining local zero waste groups or online forums where you can exchange tips, share experiences, and celebrate milestones with others on a similar journey. Having a support network can make the process more enjoyable and rewarding.

Flexibility is also crucial when setting personal goals for waste reduction. Life is unpredictable, and circumstances may change, requiring you to adjust your goals accordingly. It's important to be adaptable and open to revising your objectives as needed, without feeling discouraged. Remember that the journey towards waste reduction is a continuous process, and setbacks are a natural part of growth and learning.

Celebrating achievements, no matter how small, is vital for maintaining motivation and reinforcing positive behavior. Take time to acknowledge your progress and the impact of your efforts, whether it's reducing your waste output, discovering new sustainable practices, or inspiring others to join the movement. Celebrating successes can boost morale and provide a sense of accomplishment, fueling your commitment to further waste reduction.

Setting personal goals for waste reduction is not just about minimizing waste; it's about fostering a mindset of sustainability and conscious living. By aligning your goals with your values and priorities, you can create a lifestyle that is both fulfilling and environmentally responsible. This journey is an opportunity for personal growth, creativity, and connection with the world around you.

Chapter 2

Assessing Your Current Waste Habits

Conducting a Waste Audit

Conducting a waste audit is a crucial step in understanding your waste habits and identifying opportunities for improvement. This process involves examining the waste you generate over a specific period, providing valuable insights into your consumption patterns and the areas where you can make changes. By taking a closer look at what you discard, you can develop a more informed and strategic approach to reducing waste.

To begin a waste audit, gather the necessary materials: gloves, a tarp or large sheet, and a notebook or digital device for recording data. Choose a timeframe that suits your lifestyle, such as a week or two, to ensure you capture a comprehensive snapshot of your waste habits. During this period, collect all the waste you generate, including items from your kitchen, bathroom, office, and any other areas of your home. Be sure to include recyclables and compostable materials, as these are important components of your overall waste footprint.

Once your collection period is complete, find a suitable space to conduct the audit, such as a garage or backyard. Lay out the tarp or sheet to create a clean surface for sorting your waste. Begin by categorizing the items into different groups, such as plastics, paper, glass, metals, food waste, and miscellaneous items. This step will help you visualize the types and quantities of waste you produce, making it easier to identify patterns and areas for improvement.

As you sort through your waste, take note of any recurring items or categories that stand out. For example, you may notice a significant amount of single-use plastics, such as water bottles or packaging materials. Alternatively, you might find that food waste constitutes a large portion of your overall waste. These observations can serve as a starting point for setting specific waste reduction goals and implementing targeted strategies.

In addition to identifying problem areas, a waste audit can also highlight successes and areas where you are already making progress. For instance, you may find that you are effectively recycling paper and cardboard or that you have successfully reduced your use of disposable coffee cups. Recognizing these achievements can provide motivation and encouragement as you work towards further waste reduction.

Once you have completed the sorting and analysis, it's time to develop an action plan based on your findings. Start by setting realistic and achievable goals for reducing waste in the areas you identified as problematic. For example, if single-use plastics are a significant issue, consider investing in reusable alternatives, such as stainless steel water bottles, cloth shopping bags, and glass food storage containers. If food waste is a concern, explore strategies for meal planning, portion control, and composting to minimize waste.

Incorporate these goals into your daily routine, making small, incremental changes that can lead to lasting habits. For instance, commit to bringing reusable bags to the grocery store, packing lunches in reusable containers, or setting up a compost bin in your backyard. By integrating these practices into your everyday life, you can gradually reduce your waste footprint and contribute to a more sustainable lifestyle.

It's important to remember that conducting a waste audit is not a one-time event but an ongoing process. Regularly revisiting your waste habits and conducting periodic audits can help you track your progress, identify new areas for improvement, and adjust your strategies as needed. This iterative approach allows you to continuously refine your waste reduction efforts and stay committed to your sustainability goals.

Engaging others in the waste audit process can also enhance its effectiveness and impact. Involve family members, roommates, or colleagues in the audit, encouraging them to participate in sorting and analyzing the waste. This collaborative approach can foster a sense of shared responsibility and accountability, making it easier to implement changes and maintain momentum. Additionally, discussing your findings and strategies with others can lead to new ideas and perspectives, enriching your waste reduction journey.

Identifying Key Areas for Improvement

Embarking on a journey to reduce waste and live more sustainably begins with identifying key areas for improvement. This process involves taking a closer look at your daily habits, consumption patterns, and the systems in place within your home or workplace. By pinpointing specific areas where changes can be made, you can develop a targeted approach to waste reduction that is both effective and manageable.

The first step in identifying key areas for improvement is to conduct a thorough assessment of your current waste generation. This involves examining the types and quantities of waste you produce on a regular basis. Consider keeping a waste diary for a week or two, documenting every item you discard and the circumstances surrounding it. This exercise can provide valuable insights into your waste habits and highlight specific areas where changes can be made.

Once you have a clear understanding of your waste footprint, it's time to analyze the data and identify patterns. Look for recurring items or categories that stand out, such as single-use plastics, food waste, or packaging materials. These patterns can serve as a starting point for setting specific waste reduction goals and implementing targeted strategies. For example, if you notice a significant amount of plastic waste, you might focus on finding reusable alternatives or reducing your reliance on packaged goods.

In addition to examining your waste habits, consider the systems and processes in place within your home or workplace that contribute to waste generation. This might include your purchasing habits, storage practices, or waste disposal methods. By identifying inefficiencies or areas for improvement within these systems, you can develop more sustainable practices that reduce waste and conserve resources.

One key area for improvement is the reduction of single-use items. These products, often designed for convenience, contribute significantly to waste generation and environmental pollution. By identifying the single-use items you rely on most frequently, such as plastic bags, disposable cutlery, or coffee cups, you can explore alternatives that are more sustainable. Consider investing in reusable items, such as cloth shopping bags, stainless steel straws, or travel mugs, to replace their disposable counterparts.

Food waste is another critical area for improvement. It is estimated that a significant portion of household waste is composed of food that could have been consumed or composted. To address this issue, consider implementing strategies such as meal planning, portion control, and proper food storage. By planning meals in advance and purchasing only what you need, you can reduce the likelihood of food spoilage and waste. Additionally, composting food scraps can divert organic waste from landfills and create nutrient-rich soil for gardening.

Packaging waste is a prevalent issue in today's consumer-driven society. Many products come with excessive packaging that contributes to waste and environmental degradation. To reduce packaging waste, consider purchasing items in bulk, choosing products with minimal or recyclable packaging, and supporting companies that prioritize sustainable packaging practices. Additionally, explore opportunities to reuse or repurpose packaging materials, such as using glass jars for storage or repurposing cardboard boxes for organization.

Another area for improvement is energy consumption, which is closely linked to waste generation. By identifying energy inefficiencies within your home or workplace, you can reduce your carbon footprint and minimize waste. Consider conducting an energy audit to assess your energy usage and identify areas for improvement. Simple changes, such as switching to energy-efficient appliances, using LED light bulbs, or implementing smart thermostats, can lead to significant energy savings and waste reduction.

Water conservation is also an important aspect of waste reduction. By identifying areas where water is being wasted, such as leaks, inefficient fixtures, or excessive usage, you can implement strategies to conserve this valuable resource. Consider installing low-flow faucets and showerheads, fixing leaks promptly, and using water-saving appliances. Additionally, practice mindful water usage by turning off the tap while brushing your teeth or using a broom instead of a hose to clean outdoor spaces.

Transportation is another key area for improvement, as it contributes to both waste generation and greenhouse gas emissions. By examining your transportation habits, you can identify opportunities to reduce waste and minimize your environmental impact. Consider carpooling, using public transportation, or biking to reduce reliance on single-occupancy vehicles. Additionally, explore options for reducing emissions, such as maintaining your vehicle for optimal fuel efficiency or investing in a hybrid or electric vehicle.

Finally, consider the role of consumerism in waste generation. By examining your purchasing habits and the motivations behind them, you can identify opportunities to reduce waste and make more sustainable choices. Practice mindful consumption by prioritizing quality over quantity, supporting ethical and sustainable brands, and choosing products with a longer lifespan. Additionally, explore opportunities for borrowing, sharing, or renting items instead of purchasing new ones.

Understanding Your Waste Footprint

Understanding your waste footprint is a pivotal step in the journey towards a more sustainable lifestyle. It involves a comprehensive examination of the waste you generate, offering insights into your consumption patterns and the environmental impact of your daily choices. By gaining a clear understanding of your waste footprint, you can make informed decisions that lead to meaningful reductions in waste and a more sustainable way of living.

The concept of a waste footprint encompasses all the waste produced by an individual, household, or organization. This includes not only the visible waste that ends up in trash bins but also the hidden waste generated throughout the lifecycle of products and services. From the extraction of raw materials to manufacturing, transportation, and disposal, each stage contributes to the overall waste footprint. By considering the full lifecycle of products, you can gain a more accurate understanding of your environmental impact.

To begin assessing your waste footprint, start by examining your daily habits and routines. Consider the products you use regularly and the waste they generate. This might include packaging materials, single-use items, and food waste. Keep a waste diary for a week or two, documenting every item you discard and the circumstances surrounding it. This exercise can provide valuable insights into your waste habits and highlight specific areas where changes can be made.

In addition to examining your own waste generation, consider the broader context of your consumption patterns. This includes the energy and resources required to produce, transport, and dispose of the products you use. For example, the production of a single-use plastic bottle involves the extraction of petroleum, manufacturing, transportation, and eventual disposal, all of which contribute to your waste footprint. By understanding the full lifecycle of products, you can make more informed choices about what to consume and how to reduce waste.

One effective way to reduce your waste footprint is to prioritize the principles of the waste hierarchy: reduce, reuse, and recycle. Reducing consumption is the most effective way to minimize waste, as it prevents waste from being generated in the first place.

Consider ways to simplify your life and focus on what truly matters, such as prioritizing experiences over material possessions or choosing quality over quantity. By reducing your reliance on disposable products and unnecessary purchases, you can significantly decrease your waste footprint.

Reusing items is another powerful strategy for waste reduction. By finding new uses for items that might otherwise be discarded, you can extend their lifespan and reduce the demand for new products. This might involve repurposing glass jars for storage, using cloth bags for shopping, or donating unwanted items to charity. By embracing a mindset of creativity and resourcefulness, you can discover new ways to reuse items and reduce waste.

Recycling is an important component of waste reduction, but it should be viewed as a last resort after reducing and reusing. While recycling can help divert waste from landfills and conserve resources, it is not a perfect solution. The recycling process itself requires energy and resources, and not all materials can be recycled indefinitely. To maximize the effectiveness of recycling, be sure to follow local guidelines and separate materials properly. Additionally, consider supporting companies that prioritize recycled materials and sustainable practices.

Understanding your waste footprint also involves considering the social and environmental impacts of your consumption choices. This includes the working conditions of those involved in the production of goods, the environmental impact of resource extraction, and the effects of waste disposal on communities. By choosing products that are ethically and sustainably produced, you can reduce your waste footprint and contribute to a more equitable and sustainable world.

Engaging with your community can also play a significant role in understanding and reducing your waste footprint. By participating in local initiatives, such as community clean-ups, recycling programs, or zero waste workshops, you can connect with like-minded individuals and gain new perspectives on waste reduction. These connections can provide support, inspiration, and motivation as you work towards a more sustainable lifestyle.

Education is a powerful tool in understanding and reducing your waste footprint. By staying informed about the latest developments in sustainability, waste management, and environmental issues, you can make more informed choices and advocate for positive change. Consider attending workshops, reading books, or following reputable sources online to expand your knowledge and stay engaged with the sustainability movement.

Tools and Resources for Tracking Waste

Tracking waste effectively requires a combination of tools and resources that can help you monitor your progress, identify patterns, and make informed decisions about waste reduction. By utilizing these tools, you can gain a clearer understanding of your waste habits and develop strategies to minimize your environmental impact. This chapter will guide you through various methods and resources that can assist you in tracking waste, providing practical advice for beginners and seasoned waste reducers alike.

One of the most straightforward tools for tracking waste is a waste diary. This simple yet effective method involves recording every item you discard over a set period, such as a week or a month. By documenting the type of waste, its source, and the reason for disposal, you can gain valuable insights into your consumption patterns and identify areas for improvement. A waste diary can be kept in a notebook, on a spreadsheet, or through a digital app, depending on your preference. The key is to be consistent and honest in your recordings, as this will provide the most accurate picture of your waste habits.

Digital apps and platforms have become increasingly popular for tracking waste, offering a convenient and user-friendly way to monitor your progress. These apps often come with features such as waste categorization, goal setting, and progress tracking, making it easier to stay organized and motivated. Some apps even provide tips and resources for waste reduction, helping you discover new strategies and solutions. When choosing an app, consider factors such as ease of use, compatibility with your devices, and the specific features that align with your waste reduction goals.

In addition to digital tools, physical tools can also play a significant role in tracking waste. A kitchen scale, for example, can be used to measure the weight of food waste, providing a tangible metric for monitoring progress. By weighing food scraps before composting or disposal, you can track reductions in food waste over time and adjust your habits accordingly. Similarly, a recycling bin with compartments for different materials can help you track the volume of recyclables you generate, making it easier to identify patterns and set goals for improvement.

Community resources can also be invaluable in tracking waste and gaining support for your waste reduction efforts. Many communities offer waste audits, workshops, and educational programs that provide guidance and resources for tracking and reducing waste. Participating in these initiatives can connect you with like-minded individuals, offering opportunities for collaboration and shared learning. Additionally, local waste management facilities often provide information on recycling guidelines, composting programs, and waste reduction tips, helping you stay informed and engaged with community efforts.

Online forums and social media groups dedicated to waste reduction can serve as valuable resources for tracking waste and sharing experiences. By joining these communities, you can connect with others who are on a similar journey, exchanging tips, advice, and encouragement. These platforms often feature discussions on tracking methods, success stories, and challenges, providing a wealth of information and inspiration. Engaging with these communities can help you stay motivated and accountable, as well as provide opportunities for learning and growth.

Educational resources, such as books, articles, and documentaries, can also enhance your understanding of waste tracking and reduction. By exploring these materials, you can gain insights into the broader context of waste management, sustainability, and environmental impact. This knowledge can inform your tracking efforts, helping you make more informed decisions and develop effective strategies for waste reduction. Consider seeking out resources that align with your interests and goals, whether it's learning about the science of waste management or discovering innovative solutions for reducing waste.

Setting clear and achievable goals is an essential component of tracking waste effectively. By establishing specific objectives, you can measure your progress and stay focused on your waste reduction journey. Consider setting both short-term and long-term goals, such as reducing food waste by a certain percentage within a month or eliminating single-use plastics from your household within a year. These goals can serve as benchmarks for tracking your progress, providing motivation and a sense of accomplishment as you achieve them.

Regularly reviewing and reflecting on your tracking efforts is crucial for maintaining momentum and making continuous improvements. Set aside time each week or month to review your waste diary, app data, or other tracking tools, analyzing patterns and identifying areas for further reduction. This reflection process can help you stay accountable, celebrate successes, and adjust your strategies as needed. By maintaining a proactive and adaptive approach, you can continue to make progress towards your waste reduction goals.

Incorporating feedback from others can also enhance your tracking efforts and provide new perspectives on waste reduction. Share your tracking methods and progress with friends, family, or colleagues, inviting their input and suggestions. This collaborative approach can lead to new ideas and solutions, as well as foster a sense of shared responsibility and accountability. Additionally, consider seeking feedback from experts or professionals in the field of waste management, who can offer guidance and insights based on their experience and knowledge.

Setting Realistic and Achievable Targets

Setting realistic and achievable targets is a cornerstone of any successful waste reduction journey. These targets act as a guiding light, helping you navigate the complexities of changing habits and making sustainable choices. By establishing clear, attainable goals, you can maintain motivation, track progress, and celebrate milestones along the way.

The process begins with self-reflection and an honest assessment of your current waste habits. Take a moment to consider your daily routines and consumption patterns. What are the most significant sources of waste in your life? Is it the single-use plastics from your takeout meals, the food waste from your kitchen, or perhaps the packaging from online shopping? Identifying these areas is crucial, as it provides a foundation upon which to build your targets.

Once you've pinpointed the key areas for improvement, it's time to set specific, measurable, achievable, relevant, and time-bound (SMART) targets. These criteria ensure that your goals are not only clear and focused but also realistic and within reach. For instance, if you aim to reduce plastic waste, a SMART target might be to cut down your use of plastic bags by 50% within three months by switching to reusable alternatives.

Breaking down larger goals into smaller, manageable steps can make the process less daunting and more achievable. If your overarching goal is to reduce household waste by 30% in six months, consider focusing on one area at a time, such as the kitchen or bathroom. By tackling one aspect of waste reduction at a time, you can build momentum and confidence as you progress towards your larger goal.

Incorporating waste reduction targets into your daily routine is essential for long-term success. This might involve setting reminders to bring reusable bags when shopping, planning meals to minimize food waste, or dedicating time each week to composting. By integrating these practices into your everyday life, they become second nature, reducing the likelihood of reverting to old habits.

Accountability is a powerful tool in achieving your targets. Sharing your objectives with friends, family, or a community of like-minded individuals can provide support, encouragement, and motivation. Consider joining local zero waste groups or online forums where you can exchange tips, share experiences, and celebrate milestones with others on a similar journey. Having a support network can make the process more enjoyable and rewarding.

Flexibility is also crucial when setting targets. Life is unpredictable, and circumstances may change, requiring you to adjust your goals accordingly. It's important to be adaptable and open to revising your objectives as needed, without feeling discouraged. Remember that the journey towards waste reduction is a continuous process, and setbacks are a natural part of growth and learning.

Celebrating achievements, no matter how small, is vital for maintaining motivation and reinforcing positive behavior. Take time to acknowledge your progress and the impact of your efforts, whether it's reducing your waste output, discovering new sustainable practices, or inspiring others to join the movement. Celebrating successes can boost morale and provide a sense of accomplishment, fueling your commitment to further waste reduction.

Setting realistic and achievable targets is not just about minimizing waste; it's about fostering a mindset of sustainability and conscious living. By aligning your goals with your values and priorities, you can create a lifestyle that is both fulfilling and environmentally responsible. This journey is an opportunity for personal growth, creativity, and connection with the world around you.

Chapter 3

Reducing Waste in the Kitchen

Embracing Meal Planning and Prepping

Meal planning and prepping have emerged as powerful strategies in the quest to reduce waste and promote sustainability. By taking a proactive approach to meals, you can minimize food waste, save time, and make healthier choices. This chapter delves into the art of meal planning and prepping, offering practical advice for beginners and seasoned cooks alike.

Imagine the scene: it's a busy weekday evening, and you're standing in front of the refrigerator, unsure of what to cook. The clock is ticking, and the temptation to order takeout looms large. This scenario is all too familiar for many, leading to impulsive decisions and unnecessary waste. Meal planning offers a solution by providing a clear roadmap for your culinary week, eliminating the guesswork and reducing the likelihood of food spoilage.

The first step in embracing meal planning is to assess your schedule and dietary preferences. Consider the number of meals you'll need for the week, taking into account any social events, work commitments, or family gatherings. This assessment will help you determine the types of meals that suit your lifestyle, whether it's quick and easy dinners, hearty lunches, or nutritious breakfasts.

Once you have a clear understanding of your needs, it's time to create a meal plan. Start by selecting a variety of recipes that align with your dietary preferences and nutritional goals. Aim for a balance of proteins, carbohydrates, and vegetables, incorporating seasonal and locally sourced ingredients whenever possible. This not only supports local farmers but also reduces the carbon footprint associated with transporting food over long distances.

As you compile your meal plan, consider the concept of batch cooking. This involves preparing larger quantities of certain dishes, allowing you to enjoy leftovers throughout the week. Batch cooking can save time and energy, as well as reduce the temptation to resort to convenience foods. Dishes like soups, stews, casseroles, and grain bowls are ideal candidates for batch cooking, as they often taste even better the next day.

With your meal plan in hand, it's time to create a shopping list. A well-organized list is a crucial tool in minimizing food waste, as it helps you purchase only what you need. Before heading to the store, take inventory of your pantry, refrigerator, and freezer to avoid buying duplicates. Stick to your list while shopping, resisting the urge to make impulse purchases that may go unused.

Meal prepping is the next step in the process, involving the preparation of ingredients or entire meals in advance. This can be done on a designated day, such as Sunday, to set yourself up for a successful week. Begin by washing, chopping, and portioning ingredients, storing them in airtight containers for easy access. You might also consider cooking certain components, such as grains or proteins, to streamline meal assembly during the week.

Investing in quality storage containers is essential for effective meal prepping. Opt for containers that are durable, leak-proof, and made from sustainable materials, such as glass or stainless steel. These containers will keep your food fresh and organized, reducing the likelihood of spoilage and waste. Additionally, consider using reusable silicone bags for storing snacks or smaller portions.

As you embark on your meal planning and prepping journey, remember that flexibility is key. Life is unpredictable, and plans may need to be adjusted. Embrace the opportunity to be creative with your meals, using leftovers or surplus ingredients to create new dishes. This not only reduces waste but also adds variety and excitement to your culinary repertoire.

Involving family members or housemates in the meal planning process can enhance its effectiveness and enjoyment. Encourage them to contribute recipe ideas, assist with prepping, or take turns cooking. This collaborative approach fosters a sense of shared responsibility and can make mealtime a more enjoyable and social experience.

Meal planning and prepping offer numerous benefits beyond waste reduction. By taking control of your meals, you can save money, improve your nutritional intake, and reduce stress. The time and effort invested in planning and prepping are rewarded with the satisfaction of knowing you're making a positive impact on both your health and the environment.

As you continue to refine your meal planning and prepping skills, consider exploring new cuisines, ingredients, and cooking techniques. This culinary exploration can inspire creativity and keep your meals exciting and diverse. Additionally, stay informed about sustainable food practices, such as regenerative agriculture or plant-based diets, to further align your meal planning efforts with your sustainability goals.

Composting Turning Scraps into Soil

Composting is a transformative practice that turns everyday kitchen scraps into nutrient-rich soil, offering a sustainable solution to waste management. By embracing composting, you not only reduce the amount of waste sent to landfills but also enrich the earth, supporting plant growth and promoting a healthier ecosystem. This chapter delves into the art and science of composting, providing practical guidance for beginners eager to embark on this rewarding journey.

Imagine a world where the peels from your morning banana, the grounds from your coffee, and the trimmings from your evening salad don't end up in a landfill. Instead, they become part of a natural cycle, breaking down and returning to the earth as fertile soil. This is the magic of composting—a process that mimics nature's way of recycling organic matter.

To begin composting, it's essential to understand the basic principles that govern the process. Composting relies on the decomposition of organic materials by microorganisms, such as bacteria and fungi, which break down the matter into simpler compounds. This process requires a balance of carbon-rich materials, known as "browns," and nitrogen-rich materials, referred to as "greens." Browns include items like dried leaves, straw, and cardboard, while greens encompass kitchen scraps, grass clippings, and coffee grounds.

The first step in setting up a composting system is selecting a suitable location and container. Compost bins come in various shapes and sizes, from simple DIY setups to commercial tumblers. Choose a container that fits your space and lifestyle, ensuring it has adequate ventilation to promote airflow. If you have a backyard, consider an outdoor bin or pile, while those in apartments might opt for a compact indoor system or a worm bin, also known as vermicomposting.

Once your composting setup is in place, it's time to start adding materials. Begin with a layer of browns to provide structure and aeration, followed by a layer of greens. Continue alternating layers, maintaining a ratio of roughly three parts browns to one part greens. This balance ensures that the compost pile remains aerated and prevents odors. Avoid adding meat, dairy, or oily foods, as these can attract pests and slow down the decomposition process.

Turning the compost pile regularly is crucial for maintaining airflow and promoting even decomposition. Use a pitchfork or shovel to mix the materials every few weeks, ensuring that the outer layers are brought to the center. This aeration process introduces oxygen, which is vital for the microorganisms breaking down the organic matter. If the pile becomes too dry, add a bit of water to maintain moisture, akin to a damp sponge.

Patience is key when it comes to composting. Depending on factors such as temperature, moisture, and the materials used, the process can take anywhere from a few months to a year. You'll know your compost is ready when it resembles dark, crumbly soil and has an earthy smell. At this stage, it can be used to enrich garden beds, potted plants, or even as a top dressing for lawns.

For those without outdoor space, vermicomposting offers an excellent alternative. This method uses worms, typically red wigglers, to break down organic matter. Worm bins can be kept indoors, making them ideal for urban dwellers. The worms consume kitchen scraps and produce castings, a nutrient-rich byproduct that serves as an excellent fertilizer. Setting up a worm bin involves creating a suitable habitat with bedding materials, such as shredded newspaper or coconut coir, and adding a small amount of food scraps at a time.

Composting not only reduces waste but also offers numerous environmental benefits. By diverting organic matter from landfills, you help reduce methane emissions, a potent greenhouse gas. Additionally, compost enriches soil, improving its structure, water retention, and nutrient content. This, in turn, supports plant growth and reduces the need for chemical fertilizers, promoting a healthier ecosystem.

Engaging with your community can enhance your composting efforts and provide opportunities for learning and collaboration. Many communities offer composting workshops, shared composting facilities, or garden projects that welcome contributions of compost. Participating in these initiatives can connect you with like-minded individuals and provide valuable insights into best practices and innovative techniques.

Education is a powerful tool in promoting composting and sustainability. By sharing your knowledge and experiences with others, you can inspire them to embrace composting and make positive changes in their own lives. Consider hosting a composting demonstration, writing about your journey, or simply discussing the benefits of composting with friends and family. These conversations can spark interest and encourage others to join the movement towards a more sustainable future.

As you continue to refine your composting skills, explore new methods and materials to enhance your system. Experiment with different types of compost bins, try incorporating diverse organic materials, or delve into advanced techniques like hot composting or bokashi fermentation. This exploration can deepen your understanding of composting and provide new opportunities for growth and innovation.

Choosing Sustainable Kitchenware

The kitchen is often considered the heart of the home, a place where meals are crafted, memories are made, and conversations flow. Yet, it is also a space where significant waste can accumulate, particularly through the use of disposable or non-sustainable kitchenware. Choosing sustainable kitchenware is a crucial step in reducing your environmental footprint and fostering a more eco-friendly lifestyle. This chapter delves into the world of sustainable kitchenware, offering guidance on selecting items that are both functional and environmentally responsible.

The journey towards a sustainable kitchen begins with understanding the materials that make up your kitchenware. Traditional options like plastic, non-stick coatings, and synthetic fibers often come with environmental and health concerns. Plastic, for instance, is derived from fossil fuels and can take hundreds of years to decompose, leaching harmful chemicals into the environment along the way. Non-stick coatings, while convenient, may release toxic fumes when overheated, posing risks to both health and the environment.

In contrast, sustainable kitchenware is crafted from materials that are renewable, biodegradable, or recyclable. Bamboo, for example, is a fast-growing plant that requires minimal resources to cultivate, making it an excellent choice for cutting boards, utensils, and serving dishes. Its natural antibacterial properties and durability further enhance its appeal. Similarly, stainless steel is a highly durable and recyclable material, ideal for pots, pans, and cutlery. Its resistance to rust and corrosion ensures longevity, reducing the need for frequent replacements.

Glass is another sustainable option, particularly for food storage. Unlike plastic containers, glass does not absorb odors or stains, and it can be recycled indefinitely without losing quality. Opt for glass containers with airtight lids to keep food fresh and reduce waste. Additionally, consider using glass jars for pantry staples, such as grains, pasta, and spices, to minimize packaging waste and create an organized, visually appealing kitchen.

Cast iron cookware is a timeless choice that embodies sustainability through its longevity and versatility. A well-maintained cast iron skillet can last for generations, reducing the need for disposable or short-lived alternatives. Its ability to retain and distribute heat evenly makes it suitable for a wide range of cooking methods, from frying to baking. While cast iron requires regular seasoning to maintain its non-stick surface, this process is simple and involves applying a thin layer of oil to the cookware.

Silicone is a relatively new material in the realm of sustainable kitchenware, offering a flexible and durable alternative to plastic. It is heat-resistant, non-toxic, and available in a variety of forms, from baking mats to spatulas. Silicone's versatility and ease of cleaning make it a practical choice for eco-conscious cooks. However, it's important to choose high-quality, food-grade silicone to ensure safety and durability.

When selecting sustainable kitchenware, consider the lifecycle of each item. This includes the resources required for production, the item's durability, and its end-of-life disposal options. Prioritize products that are designed to last, reducing the need for frequent replacements and minimizing waste. Additionally, support brands and manufacturers that prioritize ethical and sustainable practices, such as fair labor conditions and environmentally friendly production methods.

Embracing second-hand or vintage kitchenware is another effective way to promote sustainability. Thrift stores, garage sales, and online marketplaces often offer a treasure trove of pre-loved kitchen items, from cast iron skillets to ceramic dishes. By giving these items a new lease on life, you reduce demand for new products and prevent usable goods from ending up in landfills.

Caring for your sustainable kitchenware is essential to ensure its longevity and effectiveness. Follow manufacturer guidelines for cleaning and maintenance, and avoid using harsh chemicals or abrasive materials that could damage the surface. For example, use gentle dish soap and a soft sponge for cleaning bamboo utensils, and regularly season cast iron cookware to maintain its non-stick properties.

Incorporating sustainable kitchenware into your daily routine can also inspire creativity and mindfulness in the kitchen. Experiment with new cooking techniques, such as using a cast iron skillet for baking bread or a bamboo steamer for preparing vegetables. These experiences can deepen your connection to the food you prepare and the tools you use, fostering a greater appreciation for the art of cooking.

As you transition to a more sustainable kitchen, remember that change doesn't have to happen overnight. Gradually replace items as they wear out or become less effective, and prioritize quality over quantity. This approach not only reduces waste but also allows you to make thoughtful, informed decisions about the products you bring into your home.

Bulk Buying and Packaging-Free Shopping

Bulk buying and packaging-free shopping are transformative practices that can significantly reduce waste and promote sustainability. These approaches not only minimize the environmental impact of excessive packaging but also offer economic benefits and encourage mindful consumption. This chapter delves into the principles and practicalities of bulk buying and packaging-free shopping, providing actionable advice for those eager to embrace a more sustainable lifestyle.

Picture a typical grocery store aisle, lined with rows of brightly colored packages, each vying for your attention. While convenient, this packaging often ends up in landfills, contributing to the growing problem of plastic pollution. Bulk buying and packaging-free shopping offer an alternative, allowing you to purchase only what you need without the excess waste.

The journey begins with understanding the concept of bulk buying. This practice involves purchasing larger quantities of items, often from bulk bins or wholesale suppliers, to reduce packaging waste and save money. Common bulk items include grains, nuts, seeds, spices, and dried fruits, but many stores also offer bulk options for liquids like oils, vinegars, and even cleaning supplies.

To get started with bulk buying, it's essential to come prepared. Bring your own reusable containers, such as glass jars, cloth bags, or stainless steel canisters, to store your purchases. Many bulk stores provide scales for weighing containers before filling them, allowing you to pay only for the product itself. This not only reduces waste but also ensures you have the exact amount you need, minimizing food waste.

Packaging-free shopping takes the concept of bulk buying a step further by eliminating packaging altogether. This approach encourages consumers to seek out products that are sold without packaging or in minimal, sustainable packaging. Farmers' markets, zero-waste stores, and co-ops are excellent places to find packaging-free options, offering fresh produce, baked goods, and other essentials without the unnecessary wrapping.

When shopping packaging-free, it's important to plan ahead. Create a shopping list based on your needs and the availability of packaging-free options in your area. This preparation helps you stay focused and avoid impulse purchases that may come with excess packaging. Additionally, consider investing in reusable produce bags, beeswax wraps, and other sustainable alternatives to single-use plastics.

One of the key benefits of bulk buying and packaging-free shopping is the potential for cost savings. By purchasing only what you need and avoiding the markup associated with branded packaging, you can often find better deals on staple items. This approach also encourages you to buy in-season produce and locally sourced goods, which are typically more affordable and environmentally friendly.

Embracing these practices can also lead to a more organized and efficient kitchen. By storing bulk items in clear, labeled containers, you can easily see what you have on hand, reducing the likelihood of overbuying or letting items go to waste. This organization not only saves time and money but also fosters a sense of mindfulness and intentionality in your consumption habits.

Community involvement can enhance your bulk buying and packaging-free shopping experience. Many communities host bulk buying clubs or co-ops, where members pool resources to purchase items in bulk at a discounted rate. Joining such groups can provide access to a wider range of products and foster a sense of camaraderie among like-minded individuals. Additionally, participating in local zero-waste initiatives or workshops can offer valuable insights and support as you navigate this lifestyle change.

Education plays a crucial role in promoting bulk buying and packaging-free shopping. By sharing your experiences and knowledge with others, you can inspire them to adopt these practices and contribute to a more sustainable future. Consider hosting a workshop, writing a blog post, or simply discussing the benefits of these approaches with friends and family. These conversations can spark interest and encourage others to join the movement towards waste reduction.

As you continue to refine your bulk buying and packaging-free shopping skills, explore new products and stores that align with your values. Experiment with different types of containers, try new recipes that incorporate bulk ingredients, and seek out local artisans or producers who prioritize sustainability. This exploration can deepen your understanding of sustainable consumption and provide new opportunities for growth and innovation.

Creative Ways to Use Leftovers

Leftovers often carry a stigma of being uninspiring or mundane, yet they hold the potential to become the stars of your culinary repertoire. By embracing creativity and resourcefulness, you can transform yesterday's meals into delightful new dishes, reducing food waste and stretching your grocery budget. This chapter explores inventive ways to breathe new life into leftovers, turning them into delicious and satisfying meals.

Consider the humble roast chicken, a staple in many households. Once the main event of a family dinner, its remnants can be repurposed into a variety of dishes. Begin by shredding the leftover meat to create a versatile base for soups, salads, or sandwiches. A comforting chicken noodle soup can be whipped up by simmering the shredded chicken with broth, vegetables, and pasta. Alternatively, toss the chicken with greens, nuts, and a tangy vinaigrette for a refreshing salad, or layer it with cheese and vegetables in a hearty sandwich.

The bones of the roast chicken should not be overlooked, as they can be used to make a rich and flavorful stock. Simply simmer the bones with water, aromatic vegetables, and herbs for several hours, then strain the liquid to create a nourishing base for future soups and stews. This homemade stock not only enhances the flavor of your dishes but also maximizes the use of the entire chicken, minimizing waste.

Leftover grains, such as rice, quinoa, or couscous, offer endless possibilities for reinvention. A simple stir-fry can be elevated by adding cooked grains to a medley of sautéed vegetables, tofu, or shrimp, seasoned with soy sauce, ginger, and garlic. For a Mediterranean twist, combine leftover grains with olives, feta, cherry tomatoes, and a drizzle of olive oil for a vibrant grain salad. These dishes are not only quick and easy to prepare but also adaptable to whatever ingredients you have on hand.

Vegetables, too, can be given a second life with a bit of creativity. Roasted or steamed vegetables can be blended into a creamy soup, seasoned with herbs and spices for added depth. Alternatively, incorporate them into a frittata or omelet, along with cheese and fresh herbs, for a satisfying breakfast or brunch. Even the odds and ends of vegetables, such as carrot tops or beet greens, can be used to make flavorful pestos or chimichurris, adding a burst of flavor to pasta, sandwiches, or grilled meats.

Stale bread is often destined for the trash, yet it can be transformed into a variety of delectable dishes. Consider making croutons by tossing cubed bread with olive oil, garlic, and herbs, then baking until golden and crispy. These croutons can add texture and flavor to salads or soups. Alternatively, use stale bread to create a comforting bread pudding, sweetened with sugar and spices, or a savory strata, layered with cheese, vegetables, and eggs.

Leftover pasta can be reinvented into a new dish with minimal effort. Toss cold pasta with a zesty dressing, fresh vegetables, and protein for a refreshing pasta salad. Alternatively, layer the pasta with sauce and cheese in a baking dish to create a hearty pasta bake, perfect for a quick weeknight dinner. These dishes not only make use of leftovers but also offer the opportunity to experiment with different flavor combinations and ingredients.

Desserts, too, can benefit from a creative approach to leftovers. Overripe fruits, such as bananas or berries, can be blended into smoothies, baked into muffins, or simmered into compotes to top yogurt or pancakes. Leftover cake or cookies can be crumbled and used as a topping for ice cream or incorporated into a trifle, layered with custard and fruit.

Involving family members or housemates in the process of reinventing leftovers can make it a fun and collaborative experience. Encourage them to contribute ideas, assist with preparation, or even take turns creating new dishes. This approach not only fosters a sense of shared responsibility but also introduces fresh perspectives and ideas into the kitchen.

As you explore creative ways to use leftovers, remember that flexibility and adaptability are key. Embrace the opportunity to experiment with new ingredients, techniques, and flavor profiles, allowing your culinary instincts to guide you. This mindset not only reduces food waste but also enhances your skills and confidence in the kitchen.

Sharing your leftover creations with others can inspire them to adopt similar practices and contribute to a more sustainable future. Consider hosting a potluck or dinner party where guests bring dishes made from leftovers, showcasing the versatility and potential of these often-overlooked ingredients. These gatherings can spark conversations about sustainability and encourage others to join the movement towards waste reduction.

Chapter 4

Sustainable Shopping Practices

The Importance of Conscious Consumerism

Conscious consumerism is a powerful movement that encourages individuals to make informed and ethical choices about the products they purchase. In a world where consumer culture often prioritizes convenience and cost over sustainability and ethics, adopting a conscious approach to consumption can have a profound impact on both the environment and society. This chapter delves into the significance of conscious consumerism, offering insights and practical advice for those seeking to align their purchasing habits with their values.

Imagine walking into a store, surrounded by an overwhelming array of products, each promising to enhance your life in some way. The choices can be daunting, yet each decision you make as a consumer carries weight. Conscious consumerism invites you to pause and consider the broader implications of your purchases, from the environmental impact of production and packaging to the ethical practices of the companies involved.

At its core, conscious consumerism is about awareness and intentionality. It involves understanding the lifecycle of a product, from raw material extraction to manufacturing, distribution, and disposal. By considering these factors, you can make choices that minimize harm to the environment and support companies that prioritize sustainability and ethical practices. This approach not only reduces your ecological footprint but also sends a powerful message to businesses about the importance of responsible production.

One of the key aspects of conscious consumerism is supporting companies that prioritize sustainability. This can involve choosing products made from renewable or recycled materials, opting for items with minimal packaging, or supporting brands that invest in sustainable practices, such as reducing carbon emissions or conserving water. By directing your purchasing power towards these companies, you encourage them to continue their efforts and inspire others to follow suit.

Ethical considerations are also central to conscious consumerism. This involves examining the labor practices of the companies you support, ensuring that workers are treated fairly and paid a living wage. Look for certifications such as Fair Trade, which guarantee that products are produced under ethical conditions. By choosing ethically produced goods, you contribute to a more equitable global economy and support the rights and well-being of workers around the world.

Conscious consumerism also extends to the realm of food, where choices can have significant environmental and social impacts. Consider the origins of the food you purchase, opting for locally sourced and organic options whenever possible. Local produce not only supports regional farmers but also reduces the carbon footprint associated with transportation. Organic farming practices, meanwhile, prioritize soil health and biodiversity, contributing to a more sustainable food system.

Reducing consumption is another important aspect of conscious consumerism. In a culture that often equates success with material accumulation, embracing minimalism can be a radical act. By focusing on quality over quantity and prioritizing experiences over possessions, you can reduce waste and live a more fulfilling life. This shift in mindset encourages you to appreciate what you have and make thoughtful decisions about new purchases.

Education plays a crucial role in conscious consumerism. By staying informed about the environmental and social impacts of different industries, you can make more informed choices and advocate for change. Consider subscribing to newsletters or following organizations that focus on sustainability and ethical consumerism. These resources can provide valuable insights and keep you updated on the latest developments and innovations in the field.

Community engagement can further enhance your efforts as a conscious consumer. Join local groups or online communities that share your values, where you can exchange ideas, resources, and support. Participating in community initiatives, such as clothing swaps or zero-waste workshops, can also provide opportunities to learn and connect with others who are committed to sustainable living.

As you navigate the world of conscious consumerism, remember that perfection is not the goal. Every small step you take towards more ethical and sustainable choices contributes to a larger movement for change. Celebrate your progress and continue to seek out new ways to align your consumption habits with your values.

Choosing Eco-Friendly Products

Choosing eco-friendly products is a vital step towards reducing your environmental impact and fostering a more sustainable lifestyle. In a world where consumer choices can significantly affect the planet, selecting products that are designed with the environment in mind can make a substantial difference. This chapter explores the principles of eco-friendly purchasing, offering practical advice for those eager to make more sustainable choices in their daily lives.

Imagine walking through a store, surrounded by shelves filled with products, each claiming to be the best choice for you and the planet. The task of discerning which items are genuinely eco-friendly can be daunting, yet it is an essential skill for the conscious consumer. Eco-friendly products are those that have been designed, manufactured, and distributed with minimal impact on the environment. This encompasses a range of factors, from the materials used and the energy consumed during production to the packaging and disposal options available.

The journey towards choosing eco-friendly products begins with understanding the materials that make up the items you purchase. Natural, renewable, and biodegradable materials are often the most sustainable choices. For instance, products made from bamboo, cork, or organic cotton are preferable to those made from synthetic materials like plastic or polyester. Bamboo, for example, is a fast-growing plant that requires minimal resources to cultivate, making it an excellent choice for a variety of products, from kitchenware to clothing.

Recycled materials also play a crucial role in eco-friendly products. Items made from recycled paper, glass, or metal help reduce the demand for virgin resources and minimize waste. When selecting products, look for labels or certifications that indicate the use of recycled content, such as the recycling symbol or specific certifications like FSC (Forest Stewardship Council) for paper products.

Energy consumption is another critical factor to consider when choosing eco-friendly products. Energy-efficient appliances, for example, use less electricity and water, reducing both your utility bills and your carbon footprint. Look for products with energy efficiency ratings, such as ENERGY STAR, which certifies that an appliance meets specific energy-saving criteria. Additionally, consider the energy used during the production and transportation of a product, opting for items that are locally sourced or produced using renewable energy whenever possible.

Packaging is often an overlooked aspect of eco-friendly purchasing, yet it can have a significant impact on the environment. Products with minimal or recyclable packaging are preferable to those with excessive or non-recyclable materials. Consider bringing your own reusable bags or containers when shopping to further reduce packaging waste. Additionally, support companies that prioritize sustainable packaging solutions, such as biodegradable materials or innovative designs that minimize waste.

The lifecycle of a product is an essential consideration in eco-friendly purchasing. This involves evaluating the durability and longevity of an item, as well as its end-of-life disposal options. Products that are designed to last, with repairable parts or modular designs, are more sustainable than disposable or short-lived alternatives. Additionally, choose items that can be easily recycled or composted at the end of their life, reducing the burden on landfills and promoting a circular economy.

Certifications and labels can be valuable tools in identifying eco-friendly products. Look for trusted certifications, such as USDA Organic, Fair Trade, or Cradle to Cradle, which indicate that a product meets specific environmental or ethical standards. These labels provide assurance that the item has been produced with consideration for the planet and its inhabitants.

Education is a powerful tool in the quest for eco-friendly purchasing. By staying informed about the environmental and social impacts of different industries, you can make more informed choices and advocate for change. Consider subscribing to newsletters or following organizations that focus on sustainability and ethical consumerism. These resources can provide valuable insights and keep you updated on the latest developments and innovations in the field.

Community engagement can further enhance your efforts to choose eco-friendly products. Join local groups or online communities that share your values, where you can exchange ideas, resources, and support. Participating in community initiatives, such as zero-waste workshops or sustainable living events, can also provide opportunities to learn and connect with others who are committed to eco-friendly living.

As you navigate the world of eco-friendly products, remember that perfection is not the goal. Every small step you take towards more sustainable choices contributes to a larger movement for change. Celebrate your progress and continue to seek out new ways to align your consumption habits with your values.

Navigating Labels and Certifications

In the modern marketplace, labels and certifications serve as vital tools for consumers seeking to make informed and ethical purchasing decisions. These symbols and seals, often found on product packaging, provide insights into the environmental, social, and health impacts of the items we buy. However, the sheer number of labels and certifications can be overwhelming, leaving many consumers unsure of what they truly signify. This chapter aims to demystify the world of labels and certifications, offering guidance on how to navigate them effectively and make choices that align with your values.

Imagine standing in the grocery aisle, faced with a myriad of products, each adorned with various labels claiming to be organic, fair trade, or sustainably sourced. The challenge lies in deciphering which of these claims are credible and which are merely marketing tactics. Understanding the meaning behind these labels is crucial for making choices that support sustainability and ethical practices.

One of the most recognized labels is "USDA Organic," which indicates that a product has been produced without synthetic fertilizers, pesticides, or genetically modified organisms (GMOs). This certification is regulated by the United States Department of Agriculture and requires rigorous compliance with organic farming standards. Choosing organic products supports agricultural practices that prioritize soil health, biodiversity, and ecological balance.

Another important certification is "Fair Trade," which ensures that products are made under fair labor conditions, with equitable pay and safe working environments for workers. Fair Trade certification also promotes sustainable farming practices and community development. By choosing Fair Trade products, you contribute to a more equitable global economy and support the rights and well-being of workers around the world.

For those concerned about the environmental impact of their purchases, the "Rainforest Alliance Certified" label is a valuable indicator. This certification ensures that products are sourced from farms or forests that adhere to sustainable practices, protecting ecosystems and wildlife while supporting the livelihoods of local communities. The Rainforest Alliance focuses on conserving biodiversity and promoting sustainable land use, making it a trusted symbol for environmentally conscious consumers.

The "Energy Star" label is a familiar sight on appliances and electronics, signifying that a product meets specific energy efficiency criteria set by the Environmental Protection Agency (EPA). Energy Star-certified products use less energy, reducing both utility bills and greenhouse gas emissions. This label is particularly important for consumers looking to minimize their carbon footprint and promote energy conservation.

In the realm of textiles, the "Global Organic Textile Standard" (GOTS) is a leading certification for organic fibers. GOTS ensures that textiles are produced using environmentally and socially responsible methods, from the harvesting of raw materials to the manufacturing and labeling of finished products. This certification covers a wide range of textiles, including clothing, bedding, and home furnishings, providing assurance that the items meet high sustainability standards.

Navigating labels and certifications also involves recognizing those that may be misleading or lack credibility. Terms like "natural" or "eco-friendly" are often used in marketing but may not be backed by specific standards or regulations. It's important to research and verify the legitimacy of such claims, seeking out third-party certifications that provide transparency and accountability.

Education is a powerful tool in understanding labels and certifications. By staying informed about the various symbols and their meanings, you can make more informed choices and advocate for change. Consider subscribing to newsletters or following organizations that focus on sustainability and ethical consumerism. These resources can provide valuable insights and keep you updated on the latest developments and innovations in the field.

Community engagement can further enhance your efforts to navigate labels and certifications. Join local groups or online communities that share your values, where you can exchange ideas, resources, and support. Participating in community initiatives, such as zero-waste workshops or sustainable living events, can also provide opportunities to learn and connect with others who are committed to eco-friendly living.

As you navigate the world of labels and certifications, remember that perfection is not the goal. Every small step you take towards more sustainable choices contributes to a larger movement for change. Celebrate your progress and continue to seek out new ways to align your consumption habits with your values.

Supporting Local and Sustainable Brands

Supporting local and sustainable brands is a meaningful way to contribute to the health of your community and the planet. By choosing to invest in businesses that prioritize ethical practices and environmental stewardship, you can help foster a more resilient and equitable economy. This chapter delves into the benefits of supporting local and sustainable brands, offering practical advice for those eager to make a positive impact through their purchasing decisions.

Imagine strolling through a bustling farmers' market, where the air is filled with the scent of fresh produce and the sounds of lively conversation. Each vendor represents a local farm or artisan, offering goods that are not only fresh and unique but also produced with care and consideration for the environment. By choosing to shop at such markets, you support local economies and reduce the carbon footprint associated with transporting goods over long distances.

Local brands often have a deep connection to their communities, sourcing materials and labor from nearby areas. This not only supports local jobs but also ensures that the economic benefits of your purchases stay within the community. When you buy from local businesses, you contribute to a cycle of reinvestment that strengthens the local economy and fosters a sense of community pride.

Sustainable brands, on the other hand, prioritize environmental responsibility in their operations. This can involve using renewable resources, minimizing waste, and reducing carbon emissions. By supporting these brands, you encourage the adoption of sustainable practices across industries, sending a powerful message about the importance of environmental stewardship.

One of the key benefits of supporting local and sustainable brands is the opportunity to access high-quality, unique products. Local artisans and producers often take great pride in their work, resulting in goods that are crafted with care and attention to detail. Whether it's a handmade piece of jewelry, a loaf of freshly baked bread, or a jar of locally sourced honey, these products offer a level of quality and authenticity that is often lacking in mass-produced items.

To begin supporting local and sustainable brands, start by exploring your community. Visit farmers' markets, craft fairs, and local shops to discover the array of products available in your area. Engage with vendors and artisans, asking questions about their sourcing and production methods. This not only provides valuable insights into the origins of the products but also fosters a sense of connection and trust between you and the producers.

Online platforms can also be a valuable resource for discovering local and sustainable brands. Many websites and apps are dedicated to connecting consumers with ethical businesses, offering a convenient way to explore options beyond your immediate vicinity. These platforms often provide detailed information about the brands' practices and values, helping you make informed decisions about your purchases.

When evaluating brands, consider their commitment to sustainability and ethical practices. Look for certifications or labels that indicate adherence to specific standards, such as Fair Trade, USDA Organic, or B Corporation. These certifications provide assurance that the brand is committed to responsible practices, from sourcing materials to treating workers fairly.

Community engagement is another powerful way to support local and sustainable brands. Join local groups or online communities that share your values, where you can exchange ideas, resources, and support. Participating in community initiatives, such as buy-local campaigns or sustainability workshops, can also provide opportunities to learn and connect with others who are committed to ethical consumption.

Advocacy plays a crucial role in promoting local and sustainable brands. By sharing your experiences and knowledge with others, you can inspire them to adopt similar practices and contribute to a more sustainable future. Consider hosting a workshop, writing a blog post, or simply discussing the benefits of supporting these brands with friends and family. These conversations can spark interest and encourage others to join the movement towards ethical consumption.

As you continue to support local and sustainable brands, explore new products and businesses that align with your values. Experiment with different types of goods, try new recipes that incorporate local ingredients, and seek out artisans or producers who prioritize sustainability. This exploration can deepen your understanding of ethical consumption and provide new opportunities for growth and innovation.

The Role of Minimalism in Zero Waste

Minimalism and zero waste are two philosophies that, when combined, create a powerful approach to living sustainably. Both advocate for reducing excess and focusing on what truly matters, yet they each bring unique perspectives and practices to the table. This chapter explores how minimalism can play a crucial role in achieving a zero-waste lifestyle, offering practical advice for those seeking to simplify their lives and reduce their environmental impact.

Picture a home filled with clutter—drawers overflowing with unused gadgets, closets packed with clothes that rarely see the light of day, and shelves lined with knick-knacks collecting dust. This scene is all too familiar in a consumer-driven society that equates success with material accumulation. Minimalism challenges this notion by encouraging individuals to pare down their possessions to only those that serve a purpose or bring joy. By embracing minimalism, you can create a living environment that is not only more serene and organized but also more sustainable.

At its core, minimalism is about intentionality. It involves making conscious choices about what you bring into your life and letting go of what no longer serves you. This mindset aligns seamlessly with the principles of zero waste, which emphasize reducing consumption and minimizing waste. By adopting a minimalist approach, you can significantly reduce the amount of waste you generate, as you become more mindful of your purchasing decisions and prioritize quality over quantity.

One of the key benefits of minimalism in the context of zero waste is the reduction of impulse buying. In a world where advertisements constantly bombard us with messages to buy more, minimalism encourages you to pause and consider whether a purchase is truly necessary. This shift in mindset not only reduces clutter but also decreases the demand for new products, ultimately leading to less waste. By focusing on what you truly need, you can make more sustainable choices and avoid contributing to the cycle of overconsumption.

Minimalism also promotes the idea of valuing experiences over possessions. Instead of accumulating material goods, minimalists often prioritize experiences that enrich their lives, such as travel, learning, or spending time with loved ones. This shift in focus can lead to a more fulfilling and meaningful life, while also reducing the environmental impact associated with producing and disposing of physical items. By choosing experiences over things, you can create lasting memories without generating waste.

Incorporating minimalism into your zero-waste journey also involves rethinking the way you approach everyday tasks. Consider the kitchen, a space where waste can quickly accumulate. By adopting a minimalist mindset, you can streamline your cooking process, using fewer ingredients and tools to create delicious meals. This not only reduces food waste but also simplifies meal preparation, making it more enjoyable and less time-consuming.

The wardrobe is another area where minimalism can have a significant impact. A minimalist approach to clothing involves curating a capsule wardrobe— a collection of versatile, high-quality pieces that can be mixed and matched to create a variety of outfits. By focusing on timeless styles and durable materials, you can reduce the need for frequent clothing purchases and minimize textile waste. This approach not only simplifies your daily routine but also encourages you to invest in pieces that align with your values and personal style.

Minimalism also extends to the digital realm, where clutter can be just as overwhelming as in the physical world. By decluttering your digital life, you can reduce the energy consumption associated with storing and managing data. This involves organizing files, unsubscribing from unnecessary emails, and limiting time spent on social media. By creating a more intentional digital environment, you can free up mental space and focus on what truly matters.

Community engagement is an essential aspect of both minimalism and zero waste. By connecting with others who share your values, you can exchange ideas, resources, and support. Participating in community initiatives, such as clothing swaps or zero-waste workshops, can also provide opportunities to learn and connect with others who are committed to sustainable living. These interactions can inspire new ways to incorporate minimalism into your life and strengthen your commitment to reducing waste.

As you embrace minimalism in your zero-waste journey, remember that perfection is not the goal. Every small step you take towards simplifying your life and reducing waste contributes to a larger movement for change. Celebrate your progress and continue to seek out new ways to align your lifestyle with your values.

Chapter 5

Zero Waste Personal Care and Hygiene

DIY Personal Care Products

Crafting your own personal care products is an empowering and sustainable practice that allows you to take control of what you put on your body while reducing waste and environmental impact. In a world where commercial products often contain a myriad of synthetic ingredients and excessive packaging, creating your own alternatives can be a refreshing and rewarding endeavor. This chapter delves into the art of DIY personal care, offering practical advice and recipes for those eager to embrace a more natural and eco-friendly approach to self-care.

Imagine opening your bathroom cabinet to find an array of jars and bottles, each filled with products you've carefully crafted yourself. These creations, made from simple, natural ingredients, not only nourish your skin and hair but also reflect your commitment to sustainability. By making your own personal care items, you can tailor them to your specific needs and preferences, ensuring that each product is as effective and gentle as possible.

The journey into DIY personal care begins with understanding the ingredients you'll be working with. Natural oils, such as coconut, jojoba, and almond, serve as excellent bases for many products, providing moisture and nourishment without the use of synthetic additives. Essential oils, like lavender, tea tree, and peppermint, offer a range of therapeutic benefits and can be used to add fragrance and enhance the properties of your creations. Other common ingredients include shea butter, beeswax, and aloe vera, each bringing unique qualities to your formulations.

One of the simplest and most versatile DIY personal care products is body scrub. By combining sugar or salt with a carrier oil and a few drops of essential oil, you can create a luxurious exfoliant that leaves your skin feeling soft and rejuvenated. Experiment with different combinations to find the perfect texture and scent for your preferences. For example, a mixture of brown sugar, coconut oil, and vanilla extract creates a warm, comforting scrub, while sea salt, olive oil, and eucalyptus offer a refreshing, invigorating experience.

Lip balm is another easy and satisfying product to make at home. By melting together beeswax, shea butter, and a carrier oil, you can create a nourishing balm that protects and hydrates your lips. Add a few drops of your favorite essential oil for fragrance, or a touch of natural colorant, like beetroot powder, for a hint of tint. Pour the mixture into small tins or tubes and allow it to cool and solidify before use.

For those seeking a natural alternative to commercial deodorants, a DIY version can be both effective and gentle on the skin. By combining baking soda, arrowroot powder, and coconut oil, you can create a paste that neutralizes odor and absorbs moisture. Customize the scent with essential oils, such as lavender or tea tree, which also offer antibacterial properties. Store the deodorant in a small jar and apply a pea-sized amount to your underarms as needed.

Shampoo bars are an eco-friendly alternative to liquid shampoos, eliminating the need for plastic bottles and reducing waste. To create your own, combine a gentle surfactant, like sodium cocoyl isethionate, with nourishing oils and butters, such as cocoa butter and argan oil. Add essential oils for fragrance and additional benefits, like rosemary for scalp health or chamomile for soothing properties. Press the mixture into molds and allow it to harden before use. These bars can be customized to suit different hair types and preferences, offering a personalized cleansing experience.

Facial masks are another area where DIY personal care can shine. By using ingredients like clay, honey, and yogurt, you can create masks that address a variety of skin concerns, from dryness to acne. For a hydrating mask, mix honey with mashed avocado and apply it to your face for 15 minutes before rinsing. For a purifying mask, combine bentonite clay with apple cider vinegar to create a paste that draws out impurities and tightens pores.

As you embark on your DIY personal care journey, remember that experimentation is key. Each person's skin and hair are unique, so it may take some trial and error to find the perfect formulations for your needs. Keep a journal to track your recipes and any adjustments you make, noting how each product affects your skin and hair over time. This process of discovery can be both educational and enjoyable, allowing you to develop a deeper understanding of your body's needs.

Community engagement can further enhance your DIY personal care experience. Join local workshops or online forums where you can share recipes, tips, and experiences with others who share your interest in natural and sustainable living. These interactions can provide valuable insights and inspiration, helping you refine your techniques and expand your repertoire of homemade products.

As you continue to explore the world of DIY personal care, consider the broader impact of your choices. By reducing your reliance on commercial products, you not only minimize your exposure to synthetic chemicals but also contribute to a more sustainable and eco-friendly lifestyle. Celebrate your progress and continue to seek out new ways to align your self-care routine with your values.

Sustainable Alternatives to Common Products

In a world increasingly aware of environmental challenges, finding sustainable alternatives to common products has become a crucial endeavor. The choices we make in our daily lives, from the items we use to the way we dispose of them, have significant impacts on the planet. By opting for sustainable alternatives, we can reduce our ecological footprint and contribute to a healthier environment. This chapter explores various sustainable options for everyday products, offering practical advice for those eager to make more eco-friendly choices.

Consider the humble toothbrush, a small yet essential item in our daily routine. Traditional toothbrushes are typically made from plastic, a material that takes hundreds of years to decompose. An alternative is the bamboo toothbrush, which is biodegradable and compostable. Bamboo is a fast-growing plant that requires minimal resources to cultivate, making it an excellent sustainable choice. By switching to a bamboo toothbrush, you can significantly reduce the amount of plastic waste generated in your household.

Another common product with a sustainable alternative is the plastic water bottle. Single-use plastic bottles contribute to pollution and are a major environmental concern. Reusable water bottles, made from materials like stainless steel or glass, offer a durable and eco-friendly solution. These bottles can be used repeatedly, reducing the need for disposable plastic and helping to conserve resources. Additionally, many reusable bottles are designed to keep beverages hot or cold, providing added convenience.

In the kitchen, plastic wrap is a staple for preserving food, yet it is neither recyclable nor biodegradable. Beeswax wraps present a sustainable alternative, made from cotton fabric coated with beeswax, resin, and jojoba oil. These wraps are reusable, washable, and compostable, making them an excellent choice for reducing plastic waste. They can be used to cover bowls, wrap sandwiches, or store produce, offering versatility and sustainability in one package.

Cleaning products are another area where sustainable alternatives can make a significant impact. Many conventional cleaning products contain harsh chemicals that can harm the environment and human health. Eco-friendly cleaning solutions, made from natural ingredients like vinegar, baking soda, and essential oils, provide a safer and more sustainable option. These products are effective at cleaning and disinfecting while being gentle on the planet. By making your own cleaning solutions or purchasing from brands that prioritize sustainability, you can reduce your exposure to harmful chemicals and minimize environmental harm.

Personal care products, such as shampoo and conditioner, often come in plastic bottles and contain synthetic ingredients. Shampoo bars and conditioner bars offer a sustainable alternative, eliminating the need for plastic packaging and reducing waste. These solid bars are made from natural ingredients and can last as long as multiple bottles of liquid shampoo, providing an eco-friendly and economical option. By choosing bars over bottles, you can reduce your plastic consumption and support brands that prioritize sustainability.

In the realm of fashion, fast fashion has become synonymous with waste and environmental degradation. Sustainable fashion, on the other hand, focuses on ethical production, quality materials, and timeless designs. By choosing clothing made from organic cotton, hemp, or recycled materials, you can support brands that prioritize environmental responsibility. Additionally, investing in high-quality, durable pieces reduces the need for frequent replacements, minimizing waste and promoting a more sustainable wardrobe.

For those who enjoy a cup of coffee or tea, single-use pods and tea bags contribute to significant waste. Reusable coffee pods and loose-leaf tea infusers offer sustainable alternatives, allowing you to enjoy your favorite beverages without the environmental impact. These options not only reduce waste but also provide a more flavorful and customizable experience. By choosing reusable over disposable, you can enjoy your daily rituals while supporting a more sustainable lifestyle.

In the bathroom, disposable razors are a common source of waste. Safety razors, made from stainless steel, offer a durable and sustainable alternative. These razors use replaceable blades, reducing the amount of waste generated compared to disposable options. Safety razors provide a close and comfortable shave while being gentle on the environment. By making the switch, you can reduce your plastic consumption and enjoy a more sustainable grooming routine.

The choices we make in our daily lives have far-reaching impacts on the environment. By opting for sustainable alternatives to common products, we can reduce our ecological footprint and contribute to a healthier planet. These alternatives not only offer environmental benefits but also provide opportunities for creativity and innovation. As you explore sustainable options, consider the broader impact of your choices and how they align with your values.

Community engagement can further enhance your efforts to adopt sustainable alternatives. Join local groups or online communities that share your values, where you can exchange ideas, resources, and support. Participating in community initiatives, such as zero-waste workshops or sustainability events, can also provide opportunities to learn and connect with others who are committed to eco-friendly living.

As you continue to explore sustainable alternatives, remember that perfection is not the goal. Every small step you take towards more sustainable choices contributes to a larger movement for change. Celebrate your progress and continue to seek out new ways to align your consumption habits with your values.

Reducing Plastic in the Bathroom

The bathroom, a sanctuary of daily rituals, often harbors an unexpected culprit: plastic. From shampoo bottles to toothbrushes, plastic pervades this intimate space, contributing significantly to environmental pollution. Reducing plastic in the bathroom is not only a step towards a more sustainable lifestyle but also an opportunity to embrace creativity and mindfulness in our daily routines. This chapter offers practical strategies for minimizing plastic use in the bathroom, transforming it into a haven of sustainability.

Imagine starting your day with a refreshing shower, surrounded by products that reflect your commitment to the environment. The first step in reducing plastic is to assess the items you currently use. Take stock of the plastic bottles, tubes, and containers that line your shelves. This inventory will help you identify areas where you can make sustainable swaps.

One of the most straightforward changes is switching to bar soap instead of liquid soap in plastic dispensers. Bar soaps are often packaged in minimal or recyclable materials, and they last longer than their liquid counterparts. Look for brands that use natural ingredients and avoid plastic packaging. For those who prefer liquid soap, consider purchasing a glass or stainless steel dispenser and refilling it with bulk soap from a local store.

Shampoo and conditioner bars are another excellent alternative to plastic bottles. These solid bars are compact, travel-friendly, and often made with natural ingredients. They eliminate the need for plastic packaging and can last as long as multiple bottles of liquid shampoo. Experiment with different brands and formulations to find the ones that best suit your hair type and preferences.

Toothbrushes are a staple in every bathroom, yet most are made from plastic. Bamboo toothbrushes offer a biodegradable alternative, with handles that can be composted after use. Some brands even offer toothbrushes with replaceable heads, reducing waste further. Pair your bamboo toothbrush with toothpaste tablets or powder, which come in recyclable or compostable packaging, eliminating the need for plastic tubes.

For those who shave, consider switching to a safety razor. Unlike disposable razors, which contribute to plastic waste, safety razors are made from durable materials like stainless steel and use replaceable metal blades. They provide a close shave and can last a lifetime with proper care. This investment not only reduces plastic waste but also offers a more economical shaving solution in the long run.

Feminine hygiene products are another area where plastic reduction can have a significant impact. Traditional pads and tampons often contain plastic components and come in plastic packaging. Reusable menstrual cups, cloth pads, and period underwear offer sustainable alternatives that are both eco-friendly and cost-effective. These products reduce waste and provide a more comfortable and convenient experience.

When it comes to skincare, many products come in plastic containers. Seek out brands that offer glass or metal packaging, or consider making your own skincare products using natural ingredients. DIY recipes for facial cleansers, moisturizers, and masks can be tailored to your skin type and preferences, providing a personalized and sustainable skincare routine.

Cleaning products in the bathroom often come in plastic bottles and contain harsh chemicals. Opt for eco-friendly cleaning solutions that use natural ingredients and come in sustainable packaging. Alternatively, make your own cleaning products using simple ingredients like vinegar, baking soda, and essential oils. These homemade solutions are effective, safe, and reduce the need for plastic packaging.

Storage solutions in the bathroom can also contribute to plastic reduction. Instead of plastic bins and organizers, choose alternatives made from natural materials like bamboo, wood, or metal. These options not only reduce plastic use but also add a touch of elegance and warmth to your bathroom decor.

As you implement these changes, remember that reducing plastic in the bathroom is a journey, not a destination. Start with small, manageable swaps and gradually incorporate more sustainable practices into your routine. Celebrate each step you take towards a plastic-free bathroom, and share your experiences with others to inspire them to join you on this journey.

Community engagement can further enhance your efforts to reduce plastic in the bathroom. Join local groups or online communities that share your values, where you can exchange ideas, resources, and support. Participating in community initiatives, such as plastic-free challenges or sustainability workshops, can also provide opportunities to learn and connect with others who are committed to eco-friendly living.

As you continue to explore ways to reduce plastic in the bathroom, consider the broader impact of your choices. By minimizing your reliance on plastic, you not only reduce your environmental footprint but also contribute to a more sustainable and equitable future. This journey not only empowers you to make a difference in your own life but also inspires others to do the same.

The Impact of Fast Fashion on Waste

Fast fashion, a term that has become synonymous with rapid production and consumption of clothing, has a profound impact on waste generation and environmental degradation. This industry, driven by the demand for inexpensive and trendy apparel, operates on a model that encourages frequent purchases and quick disposal. As a result, it contributes significantly to the growing problem of textile waste, which poses a serious threat to the environment. This chapter delves into the impact of fast fashion on waste, offering insights into the challenges it presents and exploring potential solutions for a more sustainable future.

Imagine walking into a store filled with racks of clothing, each piece reflecting the latest trends at remarkably low prices. The allure of fast fashion lies in its ability to offer consumers the opportunity to update their wardrobes frequently without breaking the bank. However, this convenience comes at a cost. The rapid turnover of fashion trends leads to a cycle of overproduction and overconsumption, resulting in an overwhelming amount of textile waste.

The production process of fast fashion is characterized by its speed and efficiency, often at the expense of quality and sustainability. To keep up with the demand for new styles, manufacturers rely on cheap materials and labor, producing garments that are not designed to last. These low-quality items are often discarded after only a few wears, contributing to the staggering amount of clothing that ends up in landfills each year. According to the Environmental Protection Agency, millions of tons of textile waste are generated annually, with a significant portion attributed to fast fashion.

The environmental impact of fast fashion extends beyond waste generation. The production of clothing involves the use of vast amounts of water, energy, and chemicals, contributing to pollution and resource depletion. For instance, the dyeing and finishing processes release harmful chemicals into waterways, affecting aquatic life and human health. Additionally, the cultivation of cotton, a common material in fast fashion, requires significant water and pesticide use, further exacerbating environmental concerns.

The social implications of fast fashion are equally troubling. The industry's reliance on low-cost labor often results in poor working conditions and inadequate wages for garment workers. Many of these workers are employed in developing countries, where labor laws and regulations may be less stringent. The pressure to produce clothing quickly and cheaply can lead to exploitative practices, raising ethical concerns about the true cost of fast fashion.

Addressing the impact of fast fashion on waste requires a multifaceted approach that involves consumers, manufacturers, and policymakers. As consumers, we have the power to influence the industry by making more conscious purchasing decisions. By prioritizing quality over quantity and investing in timeless, durable pieces, we can reduce the demand for fast fashion and minimize waste. Additionally, embracing a minimalist wardrobe and practicing mindful consumption can help break the cycle of overconsumption.

Manufacturers also play a crucial role in mitigating the impact of fast fashion. By adopting sustainable practices and prioritizing ethical production, brands can reduce their environmental footprint and promote a more responsible approach to fashion. This includes using eco-friendly materials, implementing recycling programs, and ensuring fair labor practices throughout the supply chain. Some brands have already begun to embrace these principles, offering sustainable collections and transparency in their operations.

Policymakers can support these efforts by implementing regulations and incentives that encourage sustainable practices within the fashion industry. This may include setting standards for waste reduction, promoting the use of recycled materials, and supporting initiatives that foster innovation in sustainable fashion. By creating a regulatory framework that prioritizes sustainability, governments can help drive meaningful change in the industry.

Education and awareness are also key components in addressing the impact of fast fashion on waste. By raising awareness about the environmental and social consequences of fast fashion, we can empower consumers to make informed choices and advocate for change. Educational campaigns, workshops, and community initiatives can provide valuable information and resources, inspiring individuals to take action and support sustainable fashion.

As we navigate the challenges posed by fast fashion, it is important to recognize the potential for positive change. The growing interest in sustainable fashion and the rise of ethical brands demonstrate a shift in consumer values and priorities. By supporting these initiatives and advocating for responsible practices, we can contribute to a more sustainable and equitable fashion industry.

Mindful Consumption of Beauty Products

The allure of beauty products is undeniable, with their promises of radiant skin, luscious hair, and transformative effects. Yet, beneath the glossy packaging and enticing claims lies a complex web of environmental and ethical considerations. Mindful consumption of beauty products involves making informed choices that align with your values, prioritize sustainability, and promote well-being. This chapter delves into the principles of mindful beauty consumption, offering practical guidance for those seeking to enhance their beauty routines while minimizing their ecological footprint.

Imagine standing in front of a vanity cluttered with an array of beauty products, each one vying for your attention. The first step towards mindful consumption is to assess your current collection. Take a moment to evaluate the products you own, considering their ingredients, packaging, and overall impact. This inventory will help you identify areas where you can make more sustainable choices and streamline your routine.

One of the key aspects of mindful beauty consumption is understanding the ingredients in your products. Many conventional beauty products contain synthetic chemicals, preservatives, and fragrances that can be harmful to both your health and the environment. By opting for products made from natural, organic ingredients, you can reduce your exposure to potentially harmful substances and support brands that prioritize sustainability. Look for certifications such as USDA Organic or Ecocert, which indicate that a product meets specific environmental and ethical standards.

Packaging is another critical consideration in mindful beauty consumption. The beauty industry is notorious for its excessive use of plastic and non-recyclable materials, contributing to significant waste. Seek out brands that use minimal, recyclable, or biodegradable packaging. Some companies offer refillable options, allowing you to reuse containers and reduce waste. By choosing products with sustainable packaging, you can minimize your environmental impact and support brands that are committed to reducing their carbon footprint.

The concept of minimalism can also be applied to your beauty routine. Instead of accumulating a vast array of products, focus on a curated selection of high-quality, multi-purpose items that meet your needs. This approach not only reduces waste but also simplifies your routine, saving time and money. For example, a tinted moisturizer with SPF can serve as a foundation, sunscreen, and moisturizer in one, streamlining your morning routine and reducing the number of products you need.

Mindful consumption extends beyond the products themselves to the brands you support. Research the companies behind your favorite beauty products, considering their values, practices, and commitment to sustainability. Support brands that prioritize ethical sourcing, fair labor practices, and environmental responsibility. Many companies are transparent about their supply chains and sustainability initiatives, providing information on their websites or packaging. By choosing to support these brands, you can contribute to a more ethical and sustainable beauty industry.

DIY beauty products offer another avenue for mindful consumption. By making your own skincare and haircare products, you can control the ingredients and packaging, ensuring that they align with your values. Simple recipes using natural ingredients like coconut oil, shea butter, and essential oils can be tailored to your specific needs and preferences. DIY beauty not only reduces waste but also allows for creativity and personalization in your routine.

Community engagement can further enhance your efforts towards mindful beauty consumption. Join local groups or online communities that share your values, where you can exchange ideas, resources, and support. Participating in community initiatives, such as beauty swaps or sustainability workshops, can also provide opportunities to learn and connect with others who are committed to eco-friendly living.

As you embark on your journey towards mindful beauty consumption, remember that progress is more important than perfection. Every small step you take towards more sustainable choices contributes to a larger movement for change. Celebrate your progress and continue to seek out new ways to align your beauty routine with your values.

Chapter 6

Eco-Friendly Home and Cleaning Solutions

Making Your Own Cleaning Products

The allure of a sparkling clean home is universal, yet the path to achieving it often involves a cocktail of commercial cleaning products laden with harsh chemicals. These products, while effective, can pose risks to both health and the environment. Making your own cleaning products offers a sustainable and safe alternative, allowing you to maintain a clean home while minimizing your ecological footprint. This chapter delves into the art of DIY cleaning solutions, providing practical guidance for those eager to embrace a more natural approach to cleanliness.

Picture a kitchen gleaming with the freshness of a homemade cleaner, its scent a soothing blend of citrus and herbs. The journey to creating your own cleaning products begins with understanding the basic ingredients that form the foundation of many effective solutions. Common household items such as vinegar, baking soda, lemon juice, and essential oils can be transformed into powerful cleaning agents with minimal effort.

Vinegar, a staple in many kitchens, is a versatile and potent cleaner. Its acidic nature makes it effective at cutting through grease, dissolving mineral deposits, and neutralizing odors. A simple solution of equal parts vinegar and water can serve as an all-purpose cleaner, suitable for countertops, windows, and floors. For those who find the scent of vinegar too strong, adding a few drops of essential oils like lavender or eucalyptus can provide a pleasant aroma while enhancing the cleaning power.

Baking soda, another common ingredient, is renowned for its abrasive properties and ability to neutralize odors. It can be used to scrub surfaces, remove stains, and freshen up fabrics. A paste made from baking soda and water is ideal for tackling tough grime on stovetops, ovens, and bathroom tiles. For an extra boost, combine baking soda with vinegar to create a fizzy reaction that can help dislodge stubborn dirt and debris.

Lemon juice, with its natural antibacterial and antiseptic properties, is a valuable addition to any DIY cleaning arsenal. Its acidity makes it effective at cutting through grease and removing stains, while its fresh scent leaves surfaces smelling clean and inviting. Lemon juice can be used on its own or combined with other ingredients to create a variety of cleaning solutions. For example, mixing lemon juice with olive oil creates a natural wood polish that nourishes and shines wooden surfaces.

Essential oils, derived from plants, offer a range of therapeutic benefits and can enhance the effectiveness of homemade cleaning products. Tea tree oil, known for its antimicrobial properties, is an excellent addition to disinfecting solutions. Lavender oil, with its calming scent, can be used to create a relaxing atmosphere while cleaning. Experiment with different combinations of essential oils to find the scents and properties that best suit your needs.

Creating your own cleaning products not only reduces your exposure to harmful chemicals but also minimizes waste. By reusing containers and opting for bulk ingredients, you can significantly reduce the amount of plastic and packaging waste generated in your household. Glass spray bottles, for instance, are a durable and eco-friendly option for storing homemade solutions. Label each bottle clearly to avoid confusion and ensure safe use.

The benefits of DIY cleaning products extend beyond environmental considerations. They offer a cost-effective alternative to commercial cleaners, allowing you to save money while maintaining a clean home. The ingredients used in homemade solutions are often inexpensive and readily available, making them accessible to a wide range of households. By investing a little time and effort, you can create a suite of cleaning products tailored to your specific needs and preferences.

As you embark on your DIY cleaning journey, consider involving family members or housemates in the process. This collaborative approach not only makes cleaning more enjoyable but also fosters a sense of shared responsibility for maintaining a sustainable home. Children, in particular, can benefit from learning about the importance of natural cleaning methods and the impact of their choices on the environment.

Community engagement can further enhance your efforts to make your own cleaning products. Join local workshops or online forums where you can share recipes, tips, and experiences with others who share your interest in natural and sustainable living. These interactions can provide valuable insights and inspiration, helping you refine your techniques and expand your repertoire of homemade solutions.

As you continue to explore the world of DIY cleaning products, remember that experimentation is key. Each home is unique, and it may take some trial and error to find the perfect formulations for your needs. Keep a journal to track your recipes and any adjustments you make, noting how each product affects different surfaces and areas of your home over time. This process of discovery can be both educational and enjoyable, allowing you to develop a deeper understanding of your cleaning preferences.

Sustainable Home Organization Tips

The quest for a well-organized home often leads us to purchase an array of storage solutions, many of which are made from plastic or other non-sustainable materials. However, organizing your home sustainably is not only possible but also rewarding, as it allows you to create a harmonious living space while minimizing your environmental impact. This chapter offers practical tips for achieving sustainable home organization, transforming your living environment into a model of efficiency and eco-friendliness.

Imagine walking into a room where every item has its place, and the space exudes a sense of calm and order.

The journey to sustainable home organization begins with decluttering. Before introducing new storage solutions, take the time to assess your belongings and determine what you truly need and use. This process involves making thoughtful decisions about what to keep, donate, or recycle, ultimately reducing the amount of unnecessary items in your home.

Once you've decluttered, consider repurposing existing items for storage. Glass jars, wooden crates, and fabric baskets can be transformed into stylish and functional organizers. For instance, glass jars can be used to store pantry staples, office supplies, or bathroom essentials, while wooden crates can serve as bookshelves or shoe racks. By repurposing items you already own, you reduce waste and avoid the need for new purchases.

When new storage solutions are necessary, opt for those made from sustainable materials. Bamboo, a fast-growing and renewable resource, is an excellent choice for shelving, drawer dividers, and storage boxes. Similarly, products made from recycled materials, such as cardboard or metal, offer eco-friendly alternatives to traditional plastic organizers. These options not only reduce your environmental footprint but also add a touch of natural beauty to your home.

Consider the principles of minimalism as you organize your space. Embrace the idea that less is more, and focus on creating a functional and aesthetically pleasing environment with fewer items. This approach not only simplifies your living space but also reduces the need for excessive storage solutions. By prioritizing quality over quantity, you can invest in durable, sustainable pieces that stand the test of time.

Labeling is a key component of effective organization, helping you maintain order and easily locate items. Instead of using plastic labels, opt for sustainable alternatives such as recycled paper tags or chalkboard labels. These options are not only eco-friendly but also customizable, allowing you to update labels as needed without generating waste.

Incorporate natural elements into your organization strategy to enhance the sustainability of your home. Plants, for example, can serve as both decorative and functional elements, improving air quality and adding a sense of tranquility to your space. Use plants to delineate different areas of a room or to create a natural barrier between spaces. Additionally, consider using natural fibers like cotton or linen for storage solutions, such as fabric bins or hanging organizers.

Community engagement can further support your efforts towards sustainable home organization. Participate in local swap events or online groups where you can exchange items with others, reducing the need for new purchases and promoting a circular economy. These interactions can provide valuable insights and inspiration, helping you refine your organizational techniques and discover new sustainable solutions.

As you implement these sustainable organization tips, remember that progress is more important than perfection. Every small step you take towards a more organized and eco-friendly home contributes to a larger movement for change. Celebrate your achievements and continue to seek out new ways to align your living environment with your values.

Energy Efficiency and Waste Reduction

The modern world, with its ever-growing demand for energy and resources, presents a unique challenge: how to balance our needs with the health of our planet. Energy efficiency and waste reduction are two powerful strategies that can help us achieve this balance, offering both environmental and economic benefits. By adopting practices that minimize energy consumption and reduce waste, we can create a more sustainable future for ourselves and generations to come. This chapter provides practical guidance on how to integrate these principles into everyday life, transforming our homes and habits into models of sustainability.

Consider the energy that powers your home, from the lights that illuminate your rooms to the appliances that make daily tasks more convenient. Each kilowatt-hour consumed contributes to greenhouse gas emissions, which drive climate change. By improving energy efficiency, we can reduce these emissions and lower our utility bills. One of the simplest ways to enhance energy efficiency is by switching to LED lighting. These bulbs use a fraction of the energy consumed by traditional incandescent bulbs and last significantly longer, reducing both energy use and waste.

Appliances are another area where energy efficiency can have a significant impact. When purchasing new appliances, look for those with the ENERGY STAR label, which indicates that they meet strict energy efficiency guidelines. These appliances use less energy to perform the same tasks, resulting in lower energy bills and reduced environmental impact. For existing appliances, regular maintenance, such as cleaning filters and coils, can improve efficiency and extend their lifespan.

Heating and cooling account for a substantial portion of household energy use. To reduce this consumption, consider investing in a programmable thermostat, which allows you to set temperatures based on your schedule. By adjusting the temperature when you're asleep or away from home, you can save energy without sacrificing comfort. Additionally, sealing drafts around windows and doors and adding insulation can help maintain a consistent indoor temperature, reducing the need for heating and cooling.

Renewable energy sources, such as solar panels, offer an opportunity to generate clean energy and further reduce reliance on fossil fuels. While the initial investment can be significant, many governments offer incentives and rebates to offset the cost. Over time, the savings on energy bills can make renewable energy a cost-effective and environmentally friendly option.

Waste reduction is another critical component of sustainability. The average person generates a significant amount of waste each year, much of which ends up in landfills, contributing to pollution and resource depletion. By adopting a zero-waste mindset, we can minimize waste and conserve resources. Start by evaluating your consumption habits and identifying areas where you can reduce, reuse, and recycle.

Reducing waste begins with mindful consumption. Before making a purchase, consider whether the item is truly necessary and if there are more sustainable alternatives available. Opt for products with minimal packaging or those made from recycled materials. When possible, buy in bulk to reduce packaging waste and save money.

Reusing items is another effective way to reduce waste. Instead of discarding items that are no longer needed, find new uses for them or donate them to others who can benefit. For example, glass jars can be repurposed for storage, and old clothing can be transformed into cleaning rags or craft materials. By extending the life of items, we can reduce the demand for new products and conserve resources.

Recycling is an essential component of waste reduction, but it's important to understand the guidelines in your area to ensure that items are properly sorted and processed. Many communities offer curbside recycling programs, while others have drop-off centers for specific materials. Familiarize yourself with the types of materials that can be recycled and make an effort to recycle as much as possible.

Composting is another effective way to reduce waste, particularly for organic materials such as food scraps and yard waste. By composting, you can divert these materials from landfills and create nutrient-rich soil for gardening. Composting can be done in a backyard bin or through community programs, making it accessible to a wide range of households.

Community involvement can amplify the impact of energy efficiency and waste reduction efforts. Join local environmental groups or participate in community initiatives that promote sustainability. These activities provide opportunities to learn from others, share resources, and collaborate on projects that benefit the entire community.

As you integrate energy efficiency and waste reduction practices into your life, remember that every small action contributes to a larger movement for change. Celebrate your progress and continue to seek out new ways to align your lifestyle with your values. By making informed and intentional choices, you can reduce your environmental impact, support ethical practices, and contribute to a more sustainable and equitable future.

The Role of Green Technology in the Home

Green technology has emerged as a beacon of hope in the quest for sustainable living, offering innovative solutions that reduce environmental impact while enhancing the quality of life. In the home, green technology plays a pivotal role in conserving resources, minimizing waste, and promoting energy efficiency. This chapter delves into the various ways green technology can be integrated into domestic settings, providing practical insights for those eager to embrace a more sustainable lifestyle.

Imagine a home where every element works in harmony with nature, from the energy that powers it to the materials that construct it. The foundation of such a home begins with energy efficiency, a cornerstone of green technology. Solar panels, for instance, harness the sun's energy to generate electricity, reducing reliance on fossil fuels and lowering utility bills. While the initial investment may seem daunting, many homeowners find that the long-term savings and environmental benefits make solar energy a worthwhile endeavor. Additionally, government incentives and rebates can help offset the cost, making solar panels more accessible to a wider audience.

Beyond solar energy, there are numerous other technologies that contribute to a home's energy efficiency. Smart thermostats, for example, allow homeowners to optimize heating and cooling systems by learning their habits and adjusting temperatures accordingly. This not only enhances comfort but also reduces energy consumption, leading to significant savings over time. Similarly, energy-efficient appliances, such as refrigerators, washing machines, and dishwashers, use advanced technology to perform tasks with minimal energy use, further contributing to a home's sustainability.

Water conservation is another critical aspect of green technology in the home. Low-flow fixtures, such as faucets, showerheads, and toilets, significantly reduce water usage without sacrificing performance. These fixtures are designed to maintain pressure while using less water, resulting in substantial savings on water bills and a reduced environmental footprint. Additionally, rainwater harvesting systems collect and store rainwater for use in irrigation, reducing the demand on municipal water supplies and promoting self-sufficiency.

The materials used in home construction and renovation also play a vital role in sustainability. Green building materials, such as bamboo flooring, recycled steel, and reclaimed wood, offer eco-friendly alternatives to traditional options. These materials are often more durable and require less energy to produce, making them a smart choice for environmentally conscious homeowners. Furthermore, using non-toxic paints and finishes can improve indoor air quality, creating a healthier living environment for occupants.

Waste reduction is another area where green technology can make a significant impact. Composting systems, for example, transform organic waste into nutrient-rich soil, reducing the amount of waste sent to landfills and providing a valuable resource for gardening. Many modern composting systems are designed for indoor use, making them suitable for homes of all sizes. Additionally, smart waste management systems can help households sort and recycle waste more efficiently, ensuring that recyclable materials are properly processed and reducing contamination.

Lighting is yet another domain where green technology shines. LED lighting, known for its energy efficiency and longevity, has become a staple in sustainable homes. These bulbs use a fraction of the energy consumed by traditional incandescent bulbs and last significantly longer, reducing both energy use and waste. Moreover, smart lighting systems allow homeowners to control lighting remotely, ensuring that lights are only used when needed and further conserving energy.

The integration of green technology in the home extends beyond individual systems and appliances. Whole-home automation systems, for instance, allow homeowners to monitor and control various aspects of their home environment, from energy use to security, through a single interface. This holistic approach not only enhances convenience but also promotes sustainability by providing real-time data and insights that can inform more efficient and eco-friendly choices.

Community engagement can amplify the impact of green technology in the home. By participating in local sustainability initiatives or joining online forums, homeowners can share experiences, resources, and knowledge with others who share their commitment to sustainable living. These interactions can provide valuable insights and inspiration, helping individuals refine their approaches and discover new technologies that align with their values.

As you integrate green technology into your home, remember that progress is more important than perfection. Every small step you take towards a more sustainable lifestyle contributes to a larger movement for change. Celebrate your achievements and continue to seek out new ways to align your living environment with your values.

Decluttering and Donating Responsibly

The art of decluttering is more than just tidying up; it's about creating a space that reflects your values and supports your well-being. In a world where consumerism often leads to overfilled closets and cluttered living spaces, the process of decluttering offers a path to simplicity and mindfulness. However, the journey doesn't end with simply removing items from your home. Responsible donating ensures that your unwanted belongings find new life and purpose, benefiting others and reducing waste. This chapter provides a comprehensive guide to decluttering and donating responsibly, transforming your home and contributing to a more sustainable world.

Picture a living room where every item has a purpose, and the space feels open and inviting. Achieving this sense of order begins with a thoughtful approach to decluttering. Start by setting clear goals for what you hope to achieve. Whether it's creating more space, reducing stress, or simplifying your lifestyle, having a clear vision will guide your decisions and keep you motivated throughout the process.

Begin by tackling one area of your home at a time, such as a closet, kitchen, or garage. This focused approach prevents overwhelm and allows you to see progress more quickly. As you sort through your belongings, ask yourself whether each item serves a purpose or brings you joy. If the answer is no, it's time to let it go. Be honest with yourself about what you truly need and use, and resist the temptation to hold onto items out of guilt or obligation.

Once you've identified items to part with, consider their potential for reuse or donation. Many items that no longer serve you may be valuable to others. Clothing, for example, can be donated to local shelters or thrift stores, where they can provide warmth and dignity to those in need. Household items, such as kitchenware, furniture, and electronics, can find new homes through donation centers or online platforms that connect donors with recipients.

When donating, it's important to ensure that items are in good condition and suitable for reuse. Take the time to clean and repair items as needed, and sort them into categories for easier distribution. This thoughtful approach not only shows respect for the recipients but also increases the likelihood that your donations will be accepted and appreciated.

Research local organizations and charities to find those that align with your values and accept the types of items you wish to donate. Many organizations have specific needs or guidelines, so it's important to reach out and confirm their requirements before dropping off donations. Some charities may even offer pickup services, making it easier to donate larger items like furniture or appliances.

In addition to traditional donation centers, consider alternative avenues for rehoming your belongings. Online platforms, such as community groups or social media marketplaces, allow you to connect directly with individuals who may benefit from your items. These platforms often facilitate exchanges within local communities, reducing the environmental impact of transportation and fostering a sense of connection and support.

For items that cannot be donated, explore recycling options to ensure they are disposed of responsibly. Electronics, for example, often contain hazardous materials that require special handling. Many communities offer e-waste recycling programs that safely process these items and recover valuable materials. Similarly, textiles that are too worn for donation can be recycled into new products, such as insulation or industrial rags.

As you declutter and donate, consider the impact of your consumption habits on the environment and society. By adopting a more mindful approach to acquiring new items, you can prevent future clutter and reduce waste. Before making a purchase, ask yourself whether the item is truly necessary and if it aligns with your values. Opt for quality over quantity, and choose products made from sustainable materials whenever possible.

Community involvement can enhance your efforts to declutter and donate responsibly. Participate in local swap events or organize a neighborhood donation drive to share resources and support those in need. These activities not only promote sustainability but also strengthen community bonds and foster a sense of collective responsibility.

As you embark on your decluttering journey, remember that progress is more important than perfection. Every item you responsibly donate or recycle contributes to a larger movement for change.

Celebrate your achievements and continue to seek out new ways to align your living environment with your values.